THE BIOCHEMISTRY OF CYTODIFFERENTIATION

BY D. E. S. TRUMAN

Department of Genetics
University of Edinburgh

A HALSTED PRESS BOOK

JOHN WILEY & SONS

NEW YORK · TORONTO

© 1974 Blackwell Scientific Publications
Osney Mead, Oxford
3 Nottingham Street, London W1
9 Forrest Road, Edinburgh
P.O. Box 9, North Balwyn, Victoria, Australia

First published 1974

Published in the U.S.A. and Canada by
Halsted Press, a Division of
John Wiley & Sons Inc, New York

Library of Congress Cataloging in Publication Data

Truman, Donald Ernest Samuel, 1936 –
The biochemistry of cytodifferentiation

A Halsted Press Book
Bibliography: p.
1. Cell differentiation. 2. Biological chemistry
I. Title
QH 607.T78   574.8'761   73-21785
ISBN 0 470 89190 4

Printed in Great Britain

THE BIOCHEMISTRY OF
CYTODIFFERENTIATION

This book is dedicated to
MY STUDENTS
who taught me so much about
the subject

CONTENTS

PREFACE

This book aims to bring together in an introductory form some of the ideas of embryology, genetics and biochemistry which are necessary to begin to understand the processes of development and differentiation, particularly in higher animals. While writing the book I have had in mind biochemists with an interest in development as well as zoologists and geneticists looking for an introduction to molecular embryology. In order to give better guidance to the literature I have tried, where possible, to give references to recent review articles rather than to the original papers. This has led to some injustice to some authors, to whom I apologize.

The relevance of biochemistry and genetics to developmental biology now seems obvious. This is a measure of the success of pioneers such as Joseph Needham and C.H.Waddington who have done so much to widen the scope of embryology so that it is now a meeting-point of many scientific disciplines. I have written this book whilst working in Professor Waddington's Epigenetics research group and, like me, the book owes much to the climate of ideas within that group. Many of my colleagues, both staff and students, have helped me by their conversation and ideas and I would especially like to record my thanks to J.C.Campbell and Ruth Clayton.

I have received generous help with reading all or parts of the manuscript from Ann Brown, Jenny Brown, J.C.Campbell, Kathleen Ramsay, K.Vasudeva Rao and G.G. Selman. I am very grateful to Margaret Perry for the electron micrographs which form Plates I and II.

I have benefited greatly from the advice of Professor H.Keir, and Dr R.Y.Thomson and from Mr R.Campbell of Blackwell Scientific Publications.

I thank all the authors and publishers who have given their permission for the reproduction of figures, the source of which is acknowledged in the text.

1

THE BIOLOGY OF
CELL DIFFERENTIATION

Microscopic examination of the various organs of higher plants and animals reveals that the cells of which they are composed may differ greatly in appearance. They vary in shape and size; in the relative proportions of nucleus and cytoplasm; in the nature of their cellular inclusions and organelles, in their staining properties and in their behaviour. On the other hand, comparison of the cells of corresponding tissues in very different species reveals marked similarities. Thus, the muscle cells found in vertebrate and invertebrate animals have much in common, as do the nerve cells of a wide range of animal species. This conservatism of a given cell type during evolution confirms the impression that we are dealing with a number of distinct types of cell. It is generally clear, for example, whether a cell is muscular, nervous or secretory. We do not usually find a graded series of cells forming a continuum between two extreme types, except when this series represents a sequence during the course of development.

The similarity of the same type of cell in different species is most strikingly demonstrated by experiments in which different cell types from two species are disaggregated into unattached cells and then set up in cell culture under conditions favourable to reaggregation. In some such experiments it has been found that muscle cells from rat and from calf will reaggregate together, while muscle cells and kidney cells from a single species do not form aggregates [160]. This illustrates the concept of **tissue specificity**; the existence of properties characteristic of a type of cell regardless of the species from which it is derived. The aggregation of cells in this instance is tissue-specific and is a behavioural property which is probably influenced by the nature of the surfaces of the cells. In subsequent chapters we shall find examples of biochemical properties of cells which are tissue-specific.

In some of the simpler forms of life the number of different types of cell that are present may be small: the sponges, for instance, are composed of only three distinct types of cell. In more complex organisms, such as the vertebrates, a single organ usually consists of several different types of cell forming a functional unit, and in the body as a whole the number of different cell types is much greater. A standard textbook of human histology may consider about one hundred different cell types, and while some of these may represent different stages in the life history of a single cell, such as the erythroblast, reticulocyte and erythrocyte in the erythroid series of cells (see Chapter 6), it is nevertheless clear that the number of discrete kinds of cell in the body of a vertebrate cannot be far from a hundred [55].

At the time in its life history immediately after fertilization even the most complex organism consists of but a single cell the **zygote**, endowed usually with a diploid set of chromosomes and with the potentiality of growth to form an adult organism. Cell division then takes place and after an early stage in which all the cells may be apparently identical, differences soon arise between various cells and we may say that differentiation has begun. *Cell differentiation, or cytodifferentiation, is the process which leads to the distinction between cell types in a body.* However, it is necessary at this point to distinguish between the relatively permanent and irreversible changes involved in **cytodifferentiation** and minor and reversible changes which are known as **modulations.** Thus, feeding will increase the glycogen content of liver cells and starvation will deplete the liver of glycogen. But we would not regard the change from the glycogen-poor state as an example of differentiation. Cytodifferentiation is a more permanent type of change. Once a cell has differentiated, for instance into a kidney cell, then it remains a kidney cell and does not change into another kind of cell. Many differentiated cells do not normally undergo further division but others, such as melanocytes or cartilage cells, do divide and their progeny also remain differentiated as melanocytes or as cartilage cells [156]. Even under the conditions of cell culture the differentiated form may be retained essentially unchanged, perhaps after as many as fifty cycles of cell division. Such stability of the differentiated state in culture has been achieved in experiments by Cahn and Cahn [21] using pigmented cells derived from the retina of the eye. Under certain of the conditions of culture these cells did lose their pigment granules, but on returning them to the original culture conditions the pigmentation returned, showing that even when the differentiation was not fully manifest it was nevertheless inherited. No other cell types appeared within the culture during this experiment.

When certain tissues are wounded the cells that remain may undergo an apparent reversal of the processes of differentiation known as **dedifferentiation.** The dedifferentiated cells then proliferate by cell division to form a mass of apparently undifferentiated cells, a **blastema.** As recovery from wounding continues these cells then redifferentiate to participate in the regeneration of the tissue. Though one cannot trace the fate of individual cells in the majority of cases the cells appear to redifferentiate into the same cell type as those that gave rise to the blastema, again showing that once differentiation has occurred it tends to be stable and to be inherited through a number of cell generations even though its outward expression may be in abeyance.

This stability of differentiation and its inheritance through cycles of cell division are important factors to remember when evaluating theories about the regulation of cell differentiation. In Chapter 4 some examples will be described in which it is possible to alter the cell type after differentiation, as in some plant cells and some forms of regeneration in animals. Though of great theoretical importance, these examples of instability are sufficiently rare not to destroy the general concept of the cellular inheritance of the state of differentiation.

Short-term and reversible changes may be superimposed upon the permanent changes of cytodifferentiation. An example of this can be seen in the changes in activity of the enzyme tryptophan oxygenase (E.C.1.13.1.12) in liver. The activity

2

TABLE I. A Classification of the Tissues of the Human Body

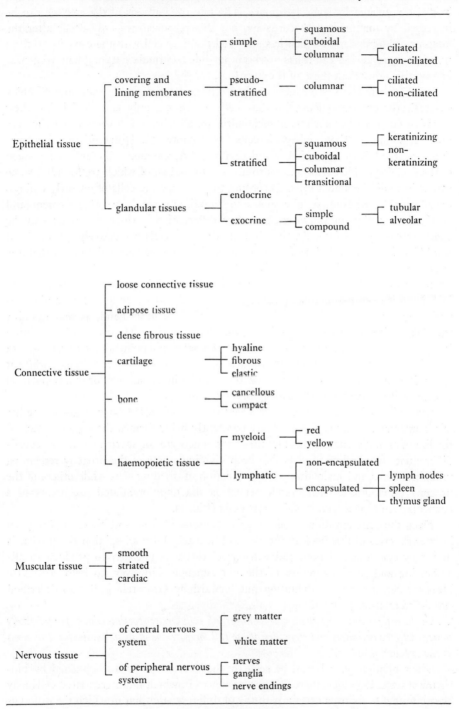

Adapted from Ham [55]

of this enzyme is affected by a number of factors, such as substrate concentration, and it is also regulated by some hormones. The rate of synthesis of the enzyme can be increased by treatment with cortisone, but this response can only occur after the long-term differentiation which has endowed the liver cell with the capacity to form tryptophan oxygenase and it has not been possible to stimulate tryptophan oxygenase in any cells other than those of the liver [72].

Cytodifferentiation is not necessarily a change from a simple state to a complex state. Erythrocytes and lens fibres are cells which not only lack nuclei, but their cytoplasmic organelles such as mitochondria are also broken down, so that they are much less complex than embryonic cells (see Chapter 6). Differentiation is sometimes regarded as specialization of function, the differentiated cell emphasizing some of the functions which are common to many cells and all of which are found to some extent in undifferentiated cells. According to this view the cells of an early embryo are all capable, for example, of response to stimuli, of some degree of movement, and of the propagation of impulses, but after differentiation these functions may be concentrated in sensory cells, muscle cells or nerve cells respectively. However, it would be almost meaningless to say that all cells synthesize haemoglobin but that it is only in erythrocytes that we are able to detect it.

Since cytodifferentiation is the origin of differences between cell types, we can look upon it from two points of view: we may study what happens during differentiation by comparing one tissue with another, or, on the other hand, we may examine the changes in an individual cell during the course of time. However, differentiation in space and time are but two aspects of the same phenomenon. Convenience in experiment may sometimes lead to the use of comparisons between tissues, while at other times it may be possible to follow the changes in a single cell or in a population of cells as it differentiates from its embryonic state.

Cytodifferentiation is not a process that is confined to the early stages in the life of an organism. Many types of cell are constantly being formed throughout much of the life of an individual, such as erythrocytes, leucocytes or spermatozoa. These cells differentiate from a line of cells that keep dividing, some of the progeny remaining undifferentiated and maintaining the population of dividing cells while others of the progeny differentiate. Cells which persist as the undifferentiated precursors of a particular cell type are known as **stem cells** (Fig. 1).

From the generalizations made in this chapter and from specific examples given in later chapters of this book the reader will probably form an opinion about what is meant by cytodifferentiation, and will realize that it is not easy to produce an all-embracing and lucid definition of the phenomenon. However, to help crystallize ideas we can use a good definition put forward by Grobstein [47] who described cytodifferentiation as including

'. . . *relatively stable, maturational changes of cellular properties which progressively concentrate the activities and structure of the cell, or portions of it, in particular directions at the expense of others.*'

Differentiation of cells can be seen by the microscope or distinguished by biochemical tests. However, there are circumstances in which undifferentiated cells may be committed to follow a certain pathway of differentiation but in which the differentiation may not yet have become distinguishable. Such cells are said to be **deter-**

4

mined. The phenomenon has been clearly demonstrated in cell culture. In the case of muscle cells growth in culture up to a certain cell density will bring about cell fusion to form multinucleate muscle fibres. But this potentiality to form fibres develops at an earlier stage, and by suitable manipulation of culture conditions the

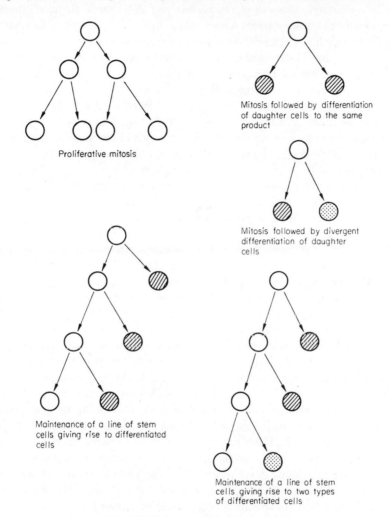

FIG. 1. Relationship Between Mitosis and Differentiation

cells may be kept in the determined but undifferentiated state for many cell generations (see Chapter 6). Similarly the capacity of cells to form cartilage may be retained for at least 35 cell generations in cells which are kept in an environment suitable for growth but not for the expression of cartilage-forming capacity [28].

Cytodifferentiation has great significance for the cell biologist and molecular biologist. It is a major process involved in the development of both higher animals and plants and poses problems which require an understanding of the mechanism of gene action. The phenomenon also has a medical significance since we may regard

5

cancer as a manifestation of pathological cell differentiation. The approach to the phenomenon which will be used in this book will be an attempt to present and to reconcile three major generalizations.

A. *The cells in an organism arise by mitosis from a fertilized egg and gradually diverge in appearance, composition, metabolism and behaviour.* Some of the many ways in which cells of the various tissues may differ will be described in Chapter 2. As yet the knowledge of the relationship between appearance and behaviour of cells on the one hand and biochemical composition and metabolism on the other remains somewhat tenuous, but the examples given in Chapter 2 are intended to illustrate the types of change which occur during cell differentiation.

B. *The properties of a cell are chiefly regulated by enzymes and other proteins, the structures of which are specified by messenger RNA's transcribed from the genes.* The ways in which enzymes may regulate cellular activity and the possible ways in which the enzymes may themselves be regulated are the subject of Chapter 3.

C. *The genes of the cells of an organism remain the same during development even though the cells' structure and metabolism alter during differentiation.* The evidence for this generalization, which is somewhat conflicting, is discussed in Chapter 4.

From these three lines of argument we see that an understanding of the phenomenon of cytodifferentiation must depend upon a comprehension of the processes whereby some genes are active in some types of cell, while a different combination of genes is active in another type. By 1934 Morgan [89] had recognized that differentiation was indeed a problem of **differential gene action** and in 1948 Spiegelman stressed that differentiation and the mechanism of gene action were but two aspects of the same problem [125]. In recent years great advances have been made in our understanding of how genes are transcribed to produce messenger RNA, and of how proteins are synthesized, with their primary amino acid sequences determining their tertiary structure and biological activity. For a clear summary of these processes the reader should turn to such reviews as those of Sargent [114], Watson [149] or Lewin [75]. In contrast to the detailed knowledge that we have of the passage of information from the gene to the protein, we must admit to great ignorance of the mechanism by which instructions are conveyed to the genes and by which some genes are activated while others are repressed. There is a further large area of ignorance of the ways in which the gene products, the proteins, actually determine the morphology and behaviour of the cell. Many developmental biologists would argue that the level of organization involved in cell differentiation is considerably more than just the production in the cytoplasm of specific concentrations of certain proteins. Others would argue that once certain polypeptide chains are synthesized they are capable of assuming the three-dimensional structure of enzymes and that these, by their specific catalytic activity, inevitably lead to the characteristic organization of the differentiated cells.

The metabolic pathways of living organisms are complex and extensively interlinked. An increase in the concentration of a single enzyme may raise the concentration of its product and lower that of its substrate. But the product of this enzyme may be the substrate of another enzyme, or might have inhibitory effects on other enzymes, while the substrate of one enzyme may also be the substrate of another so that two metabolic pathways compete with each other. Because of these types of

6

interaction it is not easy to say which of the changes occurring in differentiation are causes and which are effects. We cannot be sure which biochemical change is fundamental and which is a secondary consequence of fundamental changes.

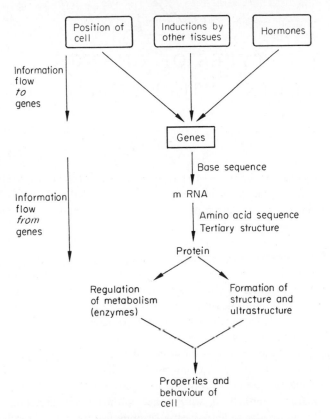

FIG. 2. Information Flow in Development

Though there is much uncertainty in the study of the biochemistry of differentiation, there is also a great challenge. There is no shortage of experimental data, as can be seen by the perusal of such compendious works as those of Needham [92], Brachet [14] or Weber [152]. Our task is to try to select the most conclusive data about the most critical stages of cytodifferentiation in order to try to produce a coherent picture of the whole process.

Summary

Cytodifferentiation is the process by which different cell types in multicellular organisms become different. It involves changes both of structure and function of the cells, brought about by changes in the chemical constituents of the cell. Since the cells of an organism are essentially similar genetically, differentiation is brought about by the activity of different sets of genes in the various cell types.

7

TYPES OF CHANGE
DURING DIFFERENTIATION

In the previous chapter the types of change which may occur during cell differentiation were mentioned briefly. In this chapter a more detailed and specific examination will be made in order to obtain a picture of the phenomenon that we are trying to explain in biochemical terms.

Cell shape
The cells of an early embryo are usually all much the same shape, though it is not unusual for them to differ in size according to how much yolk they contain. This shape is more or less spherical, with flattening in the region of the cell contacts. After differentiation the cells are much more variable in shape, often being far from spherical as elongation or flattening has taken place. Some idea of the variety of cell shapes may be gained from the drawings in Fig. 3 and from the photographs of sections of tissues in Plate 1, facing p. 10. An elongated spindle shape is found in myoblasts, and extreme elongation occurs in some neurones. A flattened disc shape is characteristic of the squamous epithelium, while other epithelial cells are columnar in form. The melanocyte is usually stellate, with a sinuous margin, while fibroblasts may be either spindle-shaped or flattened and stellate with slender tapering processes. Cell shape is a characteristic feature of some types of cell, but many other cell types could not be distinguished by their shape alone, but can only be identified when we have some idea of their composition or susceptibility to certain staining techniques.

In many differentiating cells the change of shape can be observed before there is any differentiation detectable by biochemical techniques available at present, and in some of the very important and fundamental changes in early embryonic development, such as gastrulation and the formation of the neural plate (Fig. 4) or the lens placode for example, cell elongation is one of the first discernible events [118]. Change of cell shape is often accompanied by a rearrangement of some of the organelles and inclusions within the cell. In the past attempts have been made to explain the changes in the shape of cells in terms of changes in the elasticity or growth rate of the cell membrane, but modern methods of electron microscopy have resulted in the discovery of structures which are believed to play an important part in the determination of cell shape, the **microtubules** and the **microfilaments**, which have now

8

been found in a wide variety of eukaryotic cells. The discovery of these structures is relatively recent, experimentation with them is very difficult, and there is still much controversy about the role that they play in the cell. The microtubules are apparently somewhat rigid and are usually found oriented parallel to the long axis of the cell or the elongated portion of the cell, and in some cases their appearance precedes the formation of a cell outgrowth [135]. The microfilaments are apparently contractile in nature, and seem to be related in chemical properties to the muscle protein actin. Their formation is inhibited by the drug cytochalasin B, and treatment of cells with this drug also inhibits changes in cell shape which would normally occur in cellular differentiation [154]. It appears probable that the microtubules may act as a relatively

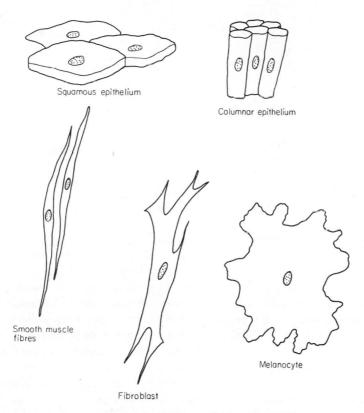

Squamous epithelium

Columnar epithelium

Smooth muscle fibres

Fibroblast

Melanocyte

FIG. 3. Variation in the Shape of Differentiated Cells

rigid skeleton within the cell, while the action of the microfilaments is to exert forces altering the shape of the cell around this skeleton. The microtubules are composed of protein subunits, each with a molecular weight of about 60,000, and these isolated subunits can be made to recombine artificially *in vitro* to form structures resembling the microtubules in the cells [127]. If these processes do represent the mechanism by which cell shape is determined, then a biochemical explanation of this important aspect of embryonic development will perhaps ultimately depend upon an understanding of the mechanisms determining the way in which microtubule and microfilament proteins aggregate together within the cell.

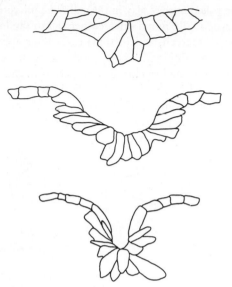

FIG. 4. The Role of Cell Shape in Morphogenesis.
   Cross-sections of the developing neural tube in *Xenopus laevis* showing changes in the shape of cells associated with the folding process. Redrawn from Schroeder, T.E.: Neurulation in *Xenopus laevis*. An analysis and model based upon light and electron microscopy. *Journal of Embryology and Experimental Morphology* **23**, 427-62 (1970).

Cell contents
One of the most dramatic ways in which cell differentiation becomes manifest is in the appearance of cellular inclusions and cellular products characteristic of the differentiated state and which may be visualized by light microscopy. The melanin granules of melanocytes, turning the cells black, vividly demonstrate the accumulation of a product specific to that cell type [156], just as the extracellular deposition of collagen distinguishes the fibroblast. Similarly we may distinguish the extracellular matrix of cartilage cells (chondrocytes), or the fat droplets which accumulate within the alveolar cells of the mammary gland, and with the light microscope we can also detect the differentiation of muscle cells by the appearance of the characteristic striations.
   With the use of histochemical stains a whole new range of processes becomes apparent, and when specific staining methods are used we are then beginning to make a biochemical analysis of cellular changes, albeit a rather crude and qualitative one. For example, the growing affinity for basic dyes which occurs as many glandular cells differentiate is a crude representation of the rising population of ribosomes in these cells as they become more active in protein synthesis.

Ultrastructure
During cell differentiation there are many changes in the fine structure of the cells which can be recognized by electron microscopy, and with the high level of resolution which has become possible with modern technique this method begins to bridge the

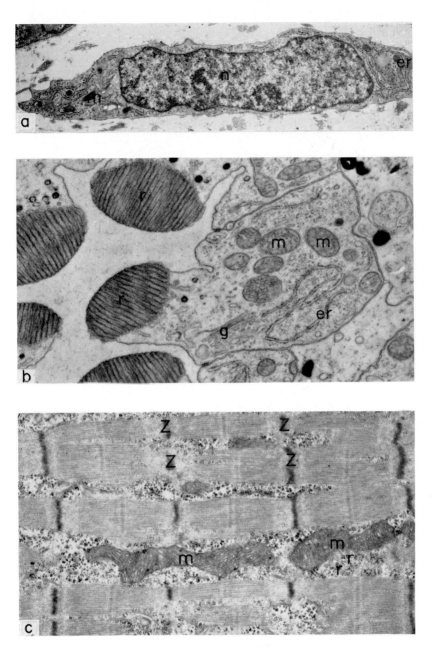

PLATE I. *Ultrastructure of Differentiated Cells*
Original electronmicrographs reproduced by kind permission of Miss M.M.Perry
(a) Fibroblast from 18-day mouse embryo. The nucleus (n) occupies a large proportion of
the cell but the remaining cytoplasm contains a rich variety of inclusions such as endoplasmic
reticulum (er) and mitochondria. ($\times$ 12 000)
(b) Retinula cell from *Drosophila* eye. Note the specialized photoreceptor region, the rhabdo-
mere (r) and the mitochondria (m), endoplasmic reticulum (er) and golgi apparatus (g).
($\times$ 13 300)
(c) Muscle from 18-day mouse embryo. The myofibrils with the longitudinal fibres of actin
and myosin are crossed by the prominent Z-bands (Z). Between them can be seen mitochondria (m) and ribosomes (r). ($\times$ 13 300)

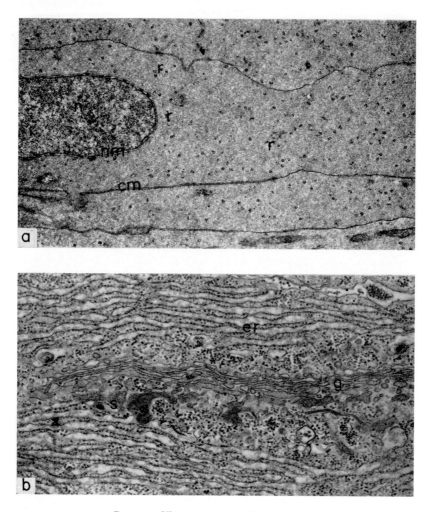

PLATE 2. *Ultrastructure of Differentiated Cells*
Original electronmicrographs reproduced by kind permission of Miss M.M.Perry
(a) Part of a lens fibre cell from 7-day chick embryo. The nucleus (n) has not yet degenerated but the cytoplasm has few inclusions, polysomes (r) and accumulated protein occupying most of the cell. The nuclear membrane (nm) and cell membrane (cm) are prominent. (×13 300)
(b) Part of the cytoplasm of a cement gland cell from a *Xenopus* larva. This secretory gland is richly endowed with rough endoplasmic reticulum (er), golgi apparatus (g) and accumulated secretory products (s). (×10 600)

gap between the biochemical level of analysis and the appearance of the cells in light microscopy. A great deal of work has been carried out in studies of the fine structure of a variety of cells. The illustrations in Plate 1 give an indication of some of the differences that can be seen among the differentiated cells of higher animals, but these can only provide a small sample. For many more examples the reader is referred to specialist works [38]. However, a list of some of the types of modification that have been found among cellular ultrastructure will serve to illustrate this aspect of cyto-differentiation.

The nucleus

The structure of the interphase nucleus varies appreciably in different types of cell, in addition to the changes which occur in the cell cycle. The nucleolus, the nuclear membrane and dense patches of **chromatin** visible by light microscopy can also be distinguished by the electron microscope. That portion of chromatin which takes up most electron-dense stain corresponds to the **heterochromatin** of the light micro-scopist, while the **euchromatin** takes up relatively little stain. The heterochromatin is believed to be inactive in **transcription**; that is to say, the DNA of this region of the chromosomes does not function as a template for RNA synthesis (see Chapter 5), and there is a correlation between increasing uptake of stain and decreasing activity in transcription as a cell becomes less active metabolically [77]. During the progression of cell types that leads to the differentiated red blood cell there are marked changes in nuclear structure: the basophilic erythroblasts have diffuse chromatin and a prominent nucleus, and the chromatin becomes progressively more condensed into large dense-staining areas as transcriptive activity is reduced, so that at the time just before the nucleus is extruded from the mammalian erythrocyte the chromatin has become entirely condensed (see Chapter 6). Similar condensation of chromatin can be seen during the differentiation of the nucleus of the spermatozoan.

The **nucleolus** undergoes considerable variation in size, reflecting its activity in the synthesis of ribosomes. It is particularly enlarged in rapidly growing cells such as those of embryos, and in cells in which protein secretion is intensive, such as the alveolar epithelium of the lactating mammary gland.

Mitochondria

The number, size and internal structure of mitochondria show appreciable variation from tissue to tissue, though the fundamental plan with an outer membrane and an inner membrane convoluted into cristae remains constant. In liver the proportion of mitochondrial volume occupied by the cristae is small, but in tissues where oxidative metabolism is rapid the mitochondria may be densely packed with cristae, as in adipose tissue or muscle [153]. Densely staining granules are frequently present in mitochondria in cells with active ion transport, such as those of the convoluted tubules of the kidney. The localization of mitochondria also varies, from being apparently random to the strict orientation found in muscle tissue, in spermatozoa and in kidney tubules.

Golgi apparatus

This organelle undergoes extensive development in secretory cells and its function appears to include the addition of carbohydrate residues to the polypeptide structure of mucoproteins and mucopolysaccharides, and the transport of such macromolecules to the outside of the cell [93]. In some secretory cells the lamellar system of the Golgi apparatus may expand to occupy a large proportion of the cell.

Endoplasmic reticulum

The structure of membranes which forms the endoplasmic reticulum varies in its development in different cell types, and it also varies in the extent to which it has ribosomes attached to it. The smooth endoplasmic reticulum is that which is devoid of ribosomes, while that portion which has attached ribosomes is the rough endoplasmic reticulum. This latter is characteristic of cells in which a high proportion of the protein which is synthesized is secreted from the cell. During differentiation the extent and type of endoplasmic reticulum frequently undergo changes. The acinar cells of the pancreas, which are active in the production of the digestive enzymes which are secreted by this organ, become packed with layers of rough endoplasmic reticulum [48]. The smooth endoplasmic reticulum is well developed in a variety of cells and it may have a number of different enzymatic functions. In the liver it is apparently involved with lipid metabolism, and in the gonads with the synthesis of steroid hormones. In the liver there is an increase in the amount of smooth endoplasmic reticulum, but not of the rough reticulum, following the injection of phenobarbitone, and this is paralleled by an increase in the activity of enzyme metabolizing the drug [38].

Lysosomes

These subcellular particles contain a variety of hydrolytic enzymes such as ribonuclease, acid phosphatases and proteases, and they can be distinguished with the electron microscope if staining techniques are used which make use of their enzymatic functions. The number of lysosomes in a cell varies greatly according to the differentiation of the cell, polymorphonuclear leucocytes and macrophages being especially rich in these organelles. Many of the processes of morphogenesis in embryos are dependent upon the action of lysosomes in bringing about the breakdown of cellular structure, an extreme case being the resorption of the tail of tadpoles at the time of metamorphosis [151].

Ribosomes

With the electron microscope it is possible to see the association of ribosomes into the functional units of protein synthesis, the **polysomes** (polyribosomes), in which a number of ribosomes are joined together while translating the same strand of messenger RNA (mRNA). At this level the picture obtained with the electron microscope represents events which can be described in molecular terms and so we can relate the ultrastructure of the cell to its function. In general the size of polysomal aggregates increases with the length of mRNA being translated and so with the molecular weight of the protein being synthesized. This is clearly seen in the polysome population of muscle cells [62].

An example of changes in polysome organization with changing differentiation can be seen as the lens fibre cell differentiates during lens regeneration in the newt.

The phenomenon of regeneration is discussed in Chapter 4. Here mention is made of but one aspect. After the dedifferentiation of the iris cell the ribosomes are either single or in very small clusters, but then a few larger aggregates become observable, though single ribosomes are still seen. As the fibre differentiates most of the ribosomes are found in clusters of a medium size, with very little variation in the number per polysome. At the end of the differentiation phase, when protein synthesis diminishes, the polysomes break down to monosomes [37].

These few examples must suffice to indicate the extent to which the ultrastructure of cells varies during differentiation. Attention has been concentrated on those organelles present in almost all cells, but it should be remembered that during differentiation new organelles may be formed, such as cilia, flagella and microvilli, while in plant cells there are chloroplasts and the structures associated with the cell wall. Moreover, by electron microscopy it is sometimes possible to detect specific substances characteristic of particular differentiated cells, such as glycogen in liver, zymogen granules in pancreas acinar cells or yolk granules in the maturing amphibian oocyte.

Cell behaviour

Cytodifferentiation is frequently accompanied by major changes in the behaviour of cells. Changes in adhesiveness are an example of this. In the early embryo and in many adult tissues the cells adhere quite firmly to each other, but in some tissues, such as blood or semen, cell adhesiveness is lost. Some non-adhesive cells, such as macrophages, also develop a capability of active movement. At an early stage of embryonic development changes in adhesiveness and the origin of cellular motility are important processes in the phenomenon of gastrulation, giving rise to the various germ layers of the embryo.

In the cells of the sensory organs the ability to respond to physical or chemical stimulation is a key aspect of differentiation. Another aspect of responsiveness which differs from one cell type to another is to be seen in the sensitivity to the influence of hormones. One of the characteristics of the muscle cell is that its ability to take up glucose is sensitive to insulin, while the liver cell is insensitive to this effect of insulin. The cells of the thyroid gland are sensitive to the hormone thyrotropin (TSH) secreted by the pituitary while other organs are unaffected. It seems probable that these changes in sensitivity to hormones may often be the result of specific modifications of the cell membrane, and it is here where we must also probably look for the changes in adhesiveness. The properties of the cell surface are clearly tissue-specific, although as yet our knowledge of the structure of the cell membrane is not sufficient for us to describe this specificity in molecular terms. It seems probable that the protein components of the cell membrane are tissue-specific in some way, though carbohydrate groups may also be involved in this specificity.

Specific substances

A characteristic of many differentiated cells is the synthesis of specific substances. Cells of the erythroid series, especially erythrocytes, are distinguished by the presence of haemoglobin, while muscle cells are characterized by a number of specific substances, including actin, myosin and myoglobin. The number of examples of tissue-specific substances which might be mentioned is very great, and it is difficult to systematize them. Table 2 mentions a few substances the synthesis of which is characteristic of particular types of cells. The substances include small molecules such as thyroxin and adrenalin, and macromolecules such as haemoglobin or crystallin. Some of the substances are retained within the cells which synthesize them, as are

TABLE 2. Some Substances Synthesized in Specific Tissues

Myoglobin	
Actin	
Myosin	Muscle
Creatine kinase (E.C.2.7.3.2)	
Creatine phosphate	
Glycogen	Muscle and liver
Serum albumin	Liver
Tryptophan oxygenase (E.C.1.13.1.12)	
Thyroxin	Thyroid
Thyroglobulin	
Chondroitin sulphate	Cartilage
Haemoglobin	Erythroid cells
Pepsin (E.C.3.4.4.1)	Gastric mucosa
Trypsin (E.C.3.4.4.4)	Pancreas acinar cells
Chymotrypsin A (E.C.3.4.4.5)	
Insulin	Pancreas islet cells
Glucagon	
Crystallins	Lens
Lactoglobulin	Mammary gland alveolar cells
UDP-galactose-glucose glycosyltransferase (E.C.2.4.1.22)	
Apatite	Bone and teeth
Melanin	Melanocytes
Retinine	Retina

actin and myosin, others are secreted to form the intercellular matrix of the tissue, such as chondroitin sulphate, while others pass into the blood stream and circulate through the body, such as insulin or serum albumin.

The development of more sensitive methods of analysis may compel us to alter some of our ideas about the tissue specificity of certain substances. For example, recent experiments have indicated the presence of proteins resembling actin in dividing cells in tissue other than muscle [104].

14

Some of the substances listed in Table 2 are proteins and so are synthesized directly under the control of particular genes, but others are synthesized as an end-product of a certain metabolic pathway, and so their presence implies activity of specific enzymes directing their synthesis. Thus not only is creatine phosphate a substance characteristic of muscle cells, but so is creatine kinase (E.C.2.7.3.2), the enzyme responsible for the synthesis of creatine phosphate.

Sometimes the detectable accumulation of specific substances in cells accompanies the microscopic or ultrastructural changes of differentiation, but when sensitive methods of assay are used, biochemical differentiation can often be found to precede structural changes. An instance of this is in the differentiation of cartilage, where the formation of chondroitin sulphate can be detected by the uptake of radioactive [35S]-sulphate before the onset of any visible differentiation of cartilage [133].

Loss of activity during differentiation

Most of the examples of changes that have been described in this chapter so far have involved the development of organelles, the synthesis of new substances or the acquisition of new patterns of behaviour. However, the differentiation of some types of cell can involve considerable loss of activity, structure or properties of behaviour. An extreme example of this is the differentiation of the erythrocyte of mammals, which involves the elimination of mitochondria, with consequent loss of the ability to perform oxidative phosphorylation, the condensation of the chromatin of the nucleus, followed by extrusion of the nucleus from the cell and the loss of capacity for cell division. At the end of this process the erythrocyte is little more than a cell membrane packed with a very high concentration of haemoglobin (see Chapter 6). The differentiation of the lens fibre involves a comparable loss of cellular ultrastructure. Even when the nucleus of the cell is not lost, the capacity for further cell division may cease with differentiation in some cells. The neurones, for instance, are incapable of division, and once myoblasts have fused to form the multinucleate myotube their nuclei no longer synthesize DNA and do not divide [60].

Importance of proteins in differentiation

In all the changes described above we may seek for the steps which are critical in that they determine the subsequent changes in the cell. From Table 2 it can be seen that many of the tissue-specific substances are proteins. Furthermore, we know that many of the non-protein substances which are characteristic of given tissues are synthesized under the influence of enzymes, which are, of course, proteins. We also saw that changes in cell shape may involve the microtubules and microfilaments which are themselves composed of protein, and it could be maintained with some plausibility that the specificity of the cell membrane is due to its protein content or to specific polysaccharides synthesized under enzymatic control.

The events which take place during the differentiation of cartilage provide us with an example of the way in which enzymes may regulate the metabolic pathways of the cell to give rise to tissue-specific substances. Cartilage is a tissue characterized by its cells, known as **chondrocytes**, being embedded in an amorphous matrix of mucopolysaccharides and collagen secreted by the cells themselves. It is this matrix which gives to the tissue its stiffness combined with a degree of flexibility which is of such mechanical importance to the organism. The process of cartilage formation, or **chondrogenesis**, has been much studied and some of the stages in the synthesis of one of the chief constituents of the mucoprotein matrix, chondroitin sulphate, are shown in Fig. 5. Some of these reactions are common to a number of tissues,

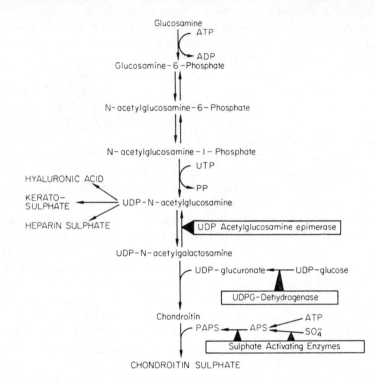

FIG. 5. Metabolic Changes Involved in Chondroitin Sulphate Formation in Cartilage
Adapted from Medoff, J.: Enzymatic events during cartilage differentiation in the embryonic limb bud. *Developmental Biology* **16**, 118-43 (1967). Copyright (1967) Academic Press, New York.

the stages in the upper part of the figure being also involved in the synthesis of hyaluronic acid, keratosulphate and heparin sulphate which are polysaccharides found in most connective tissues. The enzymes indicated in the figure are those which permit the synthesis of chondroitin sulphate. The activity of these enzymes has been studied in cultured portions of the limb in the embryonic chick by Medoff [84]. As a sensitive measure of the formation of chondroitin sulphate the uptake of radioactive [35S]-sulphate was used. The activities of the enzymes UDP acetylglucosamine epimerase (E.C.5.1.3.7), UDPG dehydrogenase (E.C.1.1.1.22) and the sulphate-activating enzymes were measured at different times after setting up the culture. The uptake of radioactive sulphate was detectable before histochemical staining with toluidine blue revealed the secretion of the matrix mucopolysaccharides. The activities are indicated in Fig. 6, which shows that the uptake of sulphate was preceded by increases in the activities of the other key enzymes in the system. For comparison

Medoff also studied the activities of two other enzymes which are not part of the pathway of chondroitin sulphate synthesis, glucose-6-phosphate dehydrogenase (E.C.1.1.1.49) and adenosine triphosphatase (E.C.3.6.1.3). The latter was

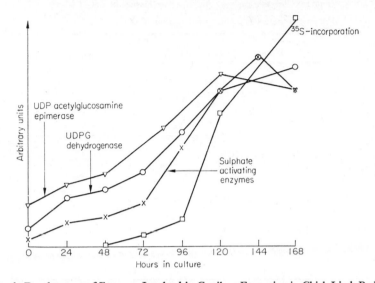

FIG. 6. Development of Enzymes Involved in Cartilage Formation in Chick Limb Buds
Adapted from Medoff, J.: Enzymatic events during cartilage differentiation in the embryonic limb bud. *Developmental Biology* **16**, 118–43 (1967). Copyright (1967) Academic Press, New York.

constant throughout the culture, while the glucose-6-phosphate dehydrogenase showed a rise of less than 30 per cent during the period of the culture.

The events of chondrogenesis exemplify the key role that certain enzymes play during differentiation and also the importance of looking for those biochemical changes which occur very early in differentiation. In the case of cartilage formation we have information about some of the crucial synthetic steps of differentiation and it has been possible to estimate the activity of the relevant enzymes. To gain an insight into the biochemistry of cytodifferentiation of particular tissues we would hope to be able to characterize some of the cell-specific substances and to measure the activity of the enzymes responsible for their synthesis.

So far in our consideration of proteins in differentiation we have considered mainly those proteins which are specific to one type of cell. Another aspect of differentiation which should be considered, however, is the importance of quantitative rather than qualitative differences between tissues. We know, for example, that many enzymes and metabolic pathways are common to a wide range of tissues: glycolysis, the citric acid cycle, oxidative phosphorylation, the hexosemonophosphate shunt, purine and pyrimidine synthesis and protein synthesis are all virtually ubiquitous in cells of higher organisms and among them must account for at least seventy enzymes, so that all these different proteins are synthesized in most cells. Nevertheless, the relative significance of these pathways differs from tissue to tissue and so some, at least, of these enzymes will differ in their relative activity. Besides these ubiquitous enzymes, there are many others which are very widespread. Table 3 lists the relative activities of a number of different enzymes in various mammalian tissues, indicating

TABLE 3. Relative Enzyme Activities in Tissues of Rat or Mouse

Enzyme activities as a percentage of that of the most active tissue

	Liver	Kidney	Spleen	Heart	Skeletal muscle	Pancreas	Brain	
Arginase E.C.3.5.3.1	100	10	1	3	1	3	1	mouse
Acetyl cholinesterase E.C.3.1.1.7	6	1	25	15	4	—	100	rat
Ca-activated ATPase	47	75	48	100	82	42	25	rat
Cytochrome oxidase E.C.1.9.3.1	30	35	11	100	16	—	32	rat
Catalase E.C.1.11.1.6	100	40	0·1	0·1	0·1	0·1	0	mouse
Leucine amino peptidase E.C.3.4.1.1	13	100	12	10	10	11	8	rat

Based on data cited by Dixon and Webb [35]

the wide range of variability of enzyme levels that may be found while more examples are shown in Fig. 7. Any theory of differentiation that we propose must be able to account for this quantitative variation among tissues.

Quantitative aspects of the regulation of metabolism have been studied in detail in connexion with carbohydrate metabolism. The various pathways by which glucose 6-phosphate may be metabolized are further considered in Chapter 3, but for the present we can note that the relative proportion metabolized by the glycolytic pathway varies from one tissue to another. For example the glycolytic pathway is highly significant in muscle, but less so in liver [120]. As seen from Fig. 8 there are many levels at which the metabolism of glucose 6-phosphate can be controlled, but clearly an increase in the activity of glucose-6-phosphate dehydrogenase (E.C.1.1.1.49) will stimulate the hexosemonophosphate shunt. Such an increase in this enzyme is found during the early cleavage stages of sea urchin embryos and is probably correlated with a growing significance of the hexosemonophosphate shunt as a source of NADPH, the reduced coenzyme required in the synthesis of fatty acids and many other important cellular constituents [4].

Isoenzymes

The changes in the relative proportions of different enzymes during differentiation are exemplified by the tissue-specific nature of the patterns of lactate dehydrogenase isoenzymes studied originally by Markert and his colleagues [79, 80]. Lactate dehydrogenase (LDH) (E.C.1.1.1.27), an enzyme with a very widespread occurrence, exists in a number of different molecular forms which can be separated by techniques such as electrophoresis. These different forms, all of which catalyse the same reaction, are known as **isoenzymes** or isozymes. There are five main isoenzymes of lactate dehydrogenase, and in many tissues all five may be present, though the relative abundance of the different forms varies from tissue to tissue. Fig. 9 illustrates the

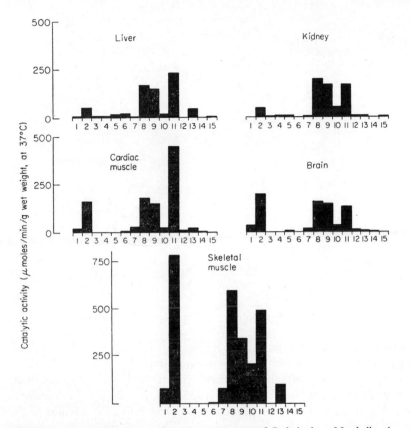

FIG. 7. Maximal Catalytic Capacities of Some Enzymes of Carbohydrate Metabolism in Different Tissues of the Rat

  1. Phosphofructokinase E.C.2.7.1.11
  2. Pyruvate kinase E.C.2.7.1.40
  3. Pyruvate carboxylase E.C.6.4.1.1
  4. Phosphopyruvate carboxylase E.C.4.1.1.32
  5. Hexosediphosphatase E.C.3.1.3.11
  6. Alanine aminotransferase E.C.2.6.1.2
  7. Fructose diphosphate aldolase E.C.4.1.2.13
  8. Glyceraldehydephosphate dehydrogenase E.C.1.2.1.12
  9. Phosphoglycerate kinase E.C.2.7.2.3
10. Phosphopyruvate hydratase E.C.4.2.1.11
11. Glucose-6-phosphatase E.C.3.1.3.9
12. Hexokinase E.C.2.7.1.1
13. Phosphoglucomutase E.C.2.7.5.1
14. UDP glucose-glycogen glycosyltransferase E.C.2.4.1.11
15. Glucose-6-phosphate dehydrogenase E.C.1.1.1.49
Drawn from data given by Scrutton and Utter [120].

patterns obtained by electrophoresis and specific staining for lactate dehydrogenase of extracts of different tissues of the mouse. It is now known that each of the isoenzymes

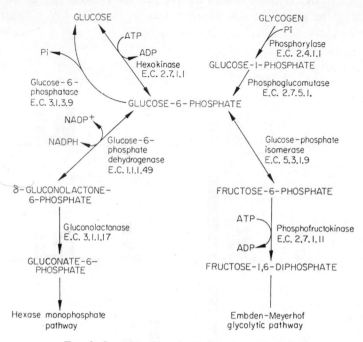

FIG. 8. Some Key Steps in Carbohydrate Metabolism

is composed of four subunits, each of which is a single polypeptide chain, and that there are two types of subunit, usually known as type A and type B. By the combination of different proportions of these subunits the five isoenzymes are built up as shown in Table 4. The two different polypeptide chains are immunologically distinct and are the products of two independent genes. The individual subunits do not have enzymatic activity, which is generated only when the tetrameric aggregates are formed. The association of subunits within the cells is apparently random, the proportion of different isoenzymes depending upon the chance aggregation of polypeptide chains. Thus if A and B subunits are present in equal numbers, the ratio of

TABLE 4. Subunit Composition of
Lactate Dehydrogenase Isoenzymes

LDH-1	B B B B
LDH-2	A B B B
LDH-3	A A B B
LDH-4	A A A B
LDH-5	A A A A

the different isoenzymes is 1 : 4 : 6 : 4 : 1, with LDH-3 (AABB) predominating, with lesser amounts of the other isoenzymes. If the ratio of subunit A to subunit B were 30 to 1, then chance aggregation gives 85 per cent of the enzyme in the form of

LDH-5, 14 per cent of LDH-4 and a trace of LDH-3. This is the situation that is found in skeletal muscle. The different isoenzymes can be generated by random reaggregation of mixtures of the subunits in a test-tube to produce results similar to those found in the tissues.

Since the different tissues have characteristic isoenzyme patterns, it follows that as differentiation occurs there must be a change in the pattern. This has often been observed, for example in the mouse kidney as shown in Fig. 9. The changing pattern

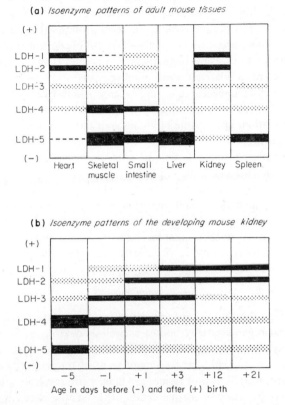

FIG. 9. Isoenzymes of Lactate Dehydrogenase Revealed by Electrophoresis
The darkness of the bands in the diagram indicates the intensity of enzyme activity revealed by staining.
Redrawn from Markert, C.L. and Ursprung, H.: The ontogeny of isozyme patterns of lactate dehydrogenase in the mouse. *Developmental Biology* **5**, 363–81 (1962). Copyright (1962) Academic Press, New York.

of the isoenzymes of different tissues, diverging from a common embryonic pattern, represents a differential accumulation of the two types of subunit, and this in turn stems from the differential activity of two genes.

Although the account of the random association of subunits given here is apparently valid in most instances, it does not always hold good and some isoenzyme patterns are not compatible with random aggregation [22]. Nevertheless, these investigations provide a good example of the way in which tissue specificity can be a consequence of variations in the concentration and proportions of proteins rather than on the presence or absence of a tissue-specific substance. In addition to lactate dehydrogenase, many other enzymes show tissue-specific isoenzyme patterns, and frequently the structure of the enzymes is also tetrameric, with four subunits in the complete enzyme molecule [81].

At this point it is perhaps advisable to mention the importance of distinguishing between enzyme concentration and enzyme activity. As will be seen in Chapter 3, the activity of enzymes is dependent upon many variables, such as the presence of inhibitors, the availability of substrates and cofactors, and sometimes on interaction with other substances in the metabolic pathway. It follows that an increase in enzymatic activity need not necessarily imply the synthesis of additional enzyme molecules. It is generally easier to measure the activity of enzymes than to estimate their concentration, and in only a relatively small number of cases have enzyme concentrations rather than activities been measured during differentiation. Changes in the localization of enzymes and changes in their stability, so that they become inactivated to varying extents during isolation, are capable of giving spurious results when one attempts to assay enzymes during cytodifferentiation. We can only infer that gene activity is represented by changes in enzyme activity when we have an indication of the actual synthesis of new enzyme molecules [144].

Enzymes are not the only proteins that may play a key role in cell differentiation: structural proteins, such as the collagen produced by fibroblasts or crystallins produced by the lens fibres, may serve as indicators of chemical differentiation, while some proteins, such as myosin, combine both structural and enzymatic functions. The synthesis and activity of proteins are clearly vital aspects of cellular differentiation.

Summary

As cells differentiate they may change in shape, in ultrastructure and behaviour, and both activities and organelles may be lost as well as developed. Sometimes specific substances are synthesized, but often cells are distinguished by a characteristic pattern of concentration of substances common to many types of cells. The differences in composition and metabolism of tissues are chiefly the result of differing activities of enzymes and so the study of the regulation of the synthesis of enzymes and other proteins is a key to understanding cytodifferentiation.

THE REGULATION OF
CELLULAR ACTIVITY

We saw in the previous chapter that the changes associated with cell differentiation included alterations in the enzyme content of cells and that these could be associated with changes in metabolism that were characteristic of the type of cell formed. In chondrogenesis, for example, we found that the production of substances specific to a cartilage was preceded by increases in activity of enzymes associated with a particular metabolic pathway (Fig. 6). On the other hand, some of the changes associated with cytodifferentiation included changes in the relative importance of metabolic pathways common to many tissues. Thus, the quantitative as well as the qualitative regulation of metabolic pathways can be of significance in cell differentiation. How does this regulation take place?

An essential characteristic of enzymes in their role in metabolic regulation is their specificity. If we look at some of the critical reactions of carbohydrate metabolism set out in Fig. 8, we can select examples of this specificity of enzymes and see the role that this phenomenon plays in directing metabolism along particular lines. Four different reactions utilizing glucose 6-phosphate are set out in Fig. 8. Two of these are isomerization reactions, not involving any other substrate. The phosphoglucomutase (E.C.2.7.5.1) reaction involves the transfer of the phosphate ester group between positions 1 and 6 of the glucose molecule, and we should note that fructose 6-phosphate is not affected by this enzyme. Similarly, glucose phosphate isomerase (E.C. 5.3.1.9) which interconverts the sugar moiety of these hexose phosphates between the aldose and the ketose forms, does not react with glucose 1-phosphate. Because of this sharp limitation upon the substrates converted, it is possible for metabolism to be directed into very different pathways by variation in the relative activities of different enzymes. Another example can be selected from Fig. 8. Glucose-6-phosphate dehydrogenase (E.C.1.1.1.49) employs as a cofactor $NADP^+$. Raising the level of $NADP^+$ stimulates this particular dehydrogenase and hence the activity of the entire hexose monophosphate shunt and leads to more glucose following this pathway of metabolism at the expense of the glycolytic pathway. The concentration of the more common dehydrogenase cofactor NAD^+ does not affect glucose-6-phosphate dehydrogenase [61].

Though many enzymes may be involved in a metabolic pathway it appears that some represent more important sites of control than others. Considerations of enzyme kinetics point to the irreversible steps of a metabolic pathway as the more likely stages for the regulation of that pathway [94]. Referring again to Fig. 8 we see that in the breakdown of glycogen by the Embden-Meyerhof glycolytic pathway the phosphorylase, phosphofructokinase and pyruvate kinase reactions are essentially irreversible and it appears that these are of special importance in the regulation of glycolysis, which is chiefly concerned with the provision of energy in the form of ATP. One function of the hexosemonophosphate shunt is probably the provision of substances involved in anabolism, the building up of cellular constituents, such as the ribose used in nucleic acid synthesis, and a further value of this pathway is the provision of reducing power in the form of NADPH, which is used in numerous biosynthetic reactions, such as fatty acid and steroid synthesis. As we saw in the previous chapter, the roles of these two pathways of glucose utilization differ in various tissues. Table 5

TABLE 5. Maximum Catalytic Capacities of Some Enzymes of Carbohydrate Metabolism in Rat Liver and Skeletal Muscle

	Catalytic capacity (μ moles/min./g. wet weight at 37°)	
	Liver	Skeletal muscle
Enzymes unique to glycolysis		
Phosphofructokinase		
E.C.2.7.1.11	3·3	80
Pyruvate kinase		
E.C.2.7.1.40	50	780
Enzymes of the pentose phosphate pathway		
Glucose-6-phosphate dehydrogenase		
E.C.1.1.1.49	6·7	0·2
Phosphogluconate dehydrogenase		
E.C.1.1.1.44	13	1·3

Based on data from Scrutton and Utter [120]

compares the relative abundance of a number of enzymes of carbohydrate metabolism in liver and skeletal muscle. Liver is a tissue much concerned with synthetic activity and one in which the hexosemonophosphate shunt is particularly significant, while muscle, a tissue which uses a great deal of ATP in the contractile process, has very active glycolysis. By comparing this table with the metabolic map in Fig. 8 we see that there is a correlation between the two, and that the key regulatory enzymes of glycolysis are much more abundant in muscle, while the two dehydrogenases that direct sugars into the hexosemonophosphate shunt are relatively concentrated in liver.

A detailed analysis of the kinetics of enzyme catalysis will be found in any good textbook of biochemistry or in a specialist work on enzymology, such as that by Dixon and Webb [35]. The rates of enzymatic reactions are influenced by temperature, pH and the ionic composition of the medium, but in the cells of higher organisms it is possible that these factors have little significance in the regulation of metabolism since they show very slight variation within most cells, and, moreover, are not capable of significant discrimination between one enzyme and another. The concentration of substrate, products and cofactors have much more specific effects on enzymatic activity. With the majority of enzymes, rising substrate concentrations lead to increased reaction rates, provided the enzyme is at a high enough concentration. When the substrate concentration is high, then the reaction rate will become dependent upon the enzyme concentration. In such cases as these, we see that one of the components of the reaction system may be of sufficiently low concentration that it becomes **rate-limiting**, while other components of the system may be of sufficiently high concentration that small changes in their concentration have no effect upon the reaction rate. In general an enzymatic system is regulated by the concentration or activity of that component which is rate-limiting. Enzyme cofactors are really a special case of the enzyme substrate, and here too the concentration of cofactor may be low enough to be rate-limiting.

The catalytic activity and specificity of enzymes are functions of their spatial molecular structure. We know that extremes of pH or temperature, or drastic chemical treatments, all of which may alter the structure of the enzymatic proteins, can destroy the catalytic power of enzymes. There are also other ways in which enzyme structures can be modified so as to influence catalytic activity. A simple and well-known instance is to be found in the digestive enzymes which are secreted in an inactive form and which become catalytic when a portion of the enzyme precursor molecule is removed:

$$\text{Trypsinogen} \rightarrow \text{Trypsin} + \text{Hexapeptide}$$

The activation of trypsinogen, pepsinogen and other zymogens is an irreversible process, but there are also important instances of enzymes which can be reversibly converted between their active and inactive forms; for example the phosphorylase of skeletal muscle which catalyses the conversion of glycogen to glucose 1-phosphate. This enzyme is converted from the inactive form known as phosphorylase b by the addition of phosphate groups to the molecule from ATP in the presence of another enzyme, phosphorylase kinase, and giving rise to the active form of phosphorylase known as phosphorylase a. The enzyme phosphorylase phosphatase catalyses the inactivation of phosphorylase a, and so the activity of phosphorylase in the muscle can be regulated by the interconversion of the active and inactive forms of the enzyme. This interconversion is affected by hormones, as will be described later, and it is through this mechanism that the hormones are able to influence glycogen utilization in muscle tissue. The activation of phosphorylase b requires, besides ATP and Mg^{++}, the enzyme phosphorylase kinase, but we find that there is a further complication because phosphorylase kinase itself exists in active and inactive forms, and its activation is in turn catalysed by an enzyme, called phosphorylase kinase-kinase, and this enzyme, yet again, has active and inactive forms. The complex cascade of reactions involved in catalysing the phosphorylation of glycogen in skeletal muscle is represented in Fig. 10. The full significance of all these steps remains at present uncertain, but clearly the regulation of enzymatic activity in higher organisms can be exceedingly complex. It is worth noting that proteins synthesized under the influence of at least three different genes are involved in the mobilization of glycogen in the skeletal muscle of mammals.

25

51842

Allosteric regulation

Another important type of variation of structure and activity of enzyme molecules is to be found in those enzymes showing the phenomenon described by Jacob, Monod and Changeux as allosteric regulation [87]. In these enzymes small molecules, often structurally unrelated to the substrate, may be bound to the enzyme at a stereospecific site which is different from the catalytically active site. These small molecules

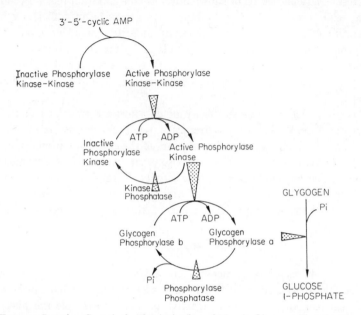

FIG. 10. Reaction Cascade for the Activation of Muscle Glycogen Phosphorylase

can either increase or decrease the catalytic activity of the enzyme, probably by influencing the spatial structure of the molecule and in particular by changing the relationship between the polypeptide subunits of the enzyme molecule. Such small molecules are known as **allosteric effectors**, and a single allosteric enzyme may be regulated by more than one effector, sometimes being inhibited by one and activated by another. In a number of instances the action of the effector regulates the cellular metabolism in a way that can clearly be seen to be advantageous, for example when glycogen synthetase is stimulated by glucose 6-phosphate and so increases the production of glycogen whenever glucose 6-phosphate begins to accumulate.

Since the concept of allosteric enzyme regulation was first put forward many examples have been discovered, but the most fully investigated of these systems are still generally found in microbial cells. It is problematic whether the major steps of regulation involved in cytodifferentiation can be attributed to allostery, the significance of which is likely to be greater in the short term, readily reversible modulations of cellular metabolism, than in the long term, almost irreversible changes of cell differentiation.

Although the activity of enzymes may be influenced by many factors, and particularly by the concentration of substrates, cofactors and other small molecules, nevertheless these small molecules are ultimately produced by enzymatic reactions. Furthermore, though enzymes may themselves be modified structurally by a variety of reactions which affect their catalytic capability, we know that in many cases these modifying reactions are themselves enzymatic. We can conclude from this that the synthesis of enzyme proteins is of major importance in the regulation of metabolic processes.

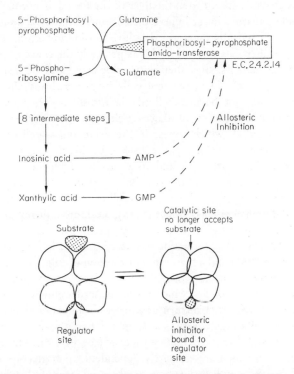

FIG. 11. An Example of Allosteric Inhibition in Purine Synthesis in the Liver and a Model of How Enzymes May Be Modified by Allosteric Effectors

The significance of the genetically regulated synthesis of enzymes in metabolic regulation is also shown by a number of diseases in man in which the synthesis of particular enzymes is defective owing to genetic factors. Accounts of some of these diseases will be found in textbooks such as that by Harris [59].

One inherited condition affecting the metabolism of aromatic amino acids is **albinism**. In the albino forms of man and animals the pigment cells fail to form melanin. This is often due to an inherited deficiency of a single enzyme, tyrosinase (o-diphenol oxidase, E.C.1.10.3.1), which is normally formed during the differentiation of melanocytes. The absence of this enzyme leads here to the absence of the end-product of the pathway, melanin, but there are not, apparently, any major consequences of the accumulation of precursors of the substrate of the enzyme or of the products of alternative metabolic pathways.

The condition known as **phenylketonuria** is associated with extreme intellectual impairment, amounting at times to idiocy. The disease was recognized by Folling in 1934, and it is characterized by the excretion of phenylpyruvic acid in the urine, while there is also a tendency to light skin and hair colouring. The inheritance of the condition is that of a Mendelian recessive character, and a high proportion of the cases known are the offspring of marriages between first cousins. Biochemical studies have shown that the disease is associated with a failure of the liver to produce the enzyme which converts phenylalanine to tyrosine, phenylalanine hydroxylase (E.C. 1.14.3.1). This enzyme, absent from the liver of the foetus, is normally found in the adult liver and so its absence in the inherited condition of phenylketonuria can be seen as an instance of a genetically determined failure in differentiation. The inability of the liver to convert phenylalanine to tyrosine is probably the cause of the pale colouring of some of these patients, since dietary tyrosine alone may be insufficient for the normal production of melanin in pigment cells by the pathway shown in Fig. 12. Since phenylalanine cannot be metabolized via tyrosine, alternative pathways are used leading to the production of substances such as phenylpyruvic acid, phenyllactic acid and phenylacetic acid, which appear in the urine and which are also probably responsible for the damage to the nervous system. In this way genetic impairment of the activity of a single enzyme can lead to a variety of biochemical consequences because of the interacting nature of the metabolic pathways of the organism.

A more complex series of metabolic events follows from a deficiency of galactose-1-phosphate uridyl transferase (E.C.2.7.7.10). **Galactosaemia** is a rare disease inherited as a Mendelian recessive. The usual symptoms are enlargement of the liver, slow growth, a tendency to the formation of opacities in the lens of the eye known as cataract, mental retardation and a high rate of excretion of galactose in the urine. The absence of the enzyme galactose-1-phosphate uridyl transferase prevents the interconversion of galactose and glucose which normally occurs chiefly in the liver. Galactose 1-phosphate, derived from dietary galactose, accumulates. This substance, an analogue of glucose 1-phosphate, acts as a competitive inhibitor of phosphoglucomutase (E.C.2.7.5.1) and thereby inhibits glycolysis (see Fig. 8). Those tissues, such as the lens of the eye, which are especially dependent upon glycolysis, are severely affected, and so the lens may become opaque. The inhibition of glycolysis means that glycogen is prone to accumulate in the liver, and this is apparently responsible for the liver enlargement that is characteristic of the disease. Again we see from this how a single enzyme defect can have a series of metabolic consequences and how some of these may lead to morphological changes, in this case an enlarged liver. As yet there are few examples of gene action in development in which the events are as well elucidated as those in galactosaemia, yet nevertheless the condition provides a model of the sort of changes that we must look for if we are to analyse events in differentiation at the molecular level.

The inherited biochemical defects in man show that the presence or absence of specific genetic activity in certain tissues is under genetic control. We cannot yet say in connexion with these examples whether the enzyme defects are due to the synthesis of an altered protein with no enzymatic activity or whether there is a total suppression of the synthesis of the enzyme. From what we know of the genetic defects in the synthesis of haemoglobin we may expect that mutations affecting enzymes in the cells would range from small changes in amino acid composition, such as the replacement of one amino acid by another, up to major changes in the structure of the protein and also complete suppression of its synthesis [105]. In order to see how these genetic changes may come about we must examine the mechanism of protein synthesis and the way in which it is controlled.

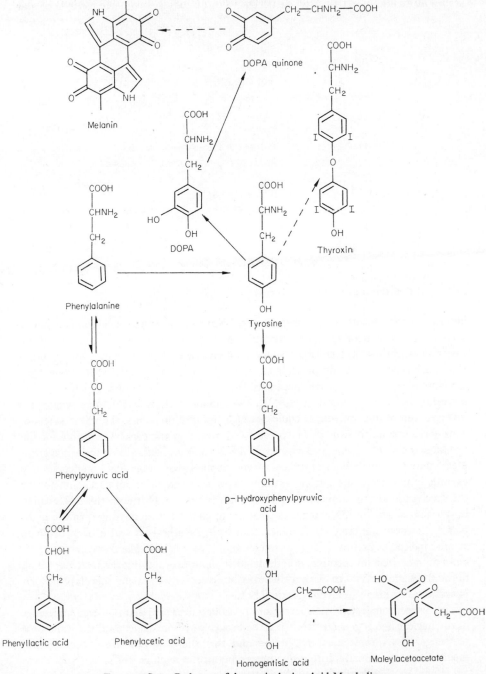

FIG. 12. Some Pathways of Aromatic Amino Acid Metabolism

The genetic regulation of protein structure

In the past twenty years or so much detailed knowledge has been applied to the construction of a model of protein synthesis and of its direction by genes. This model is now so well known that it is not necessary to discuss the evidence for it here, nor

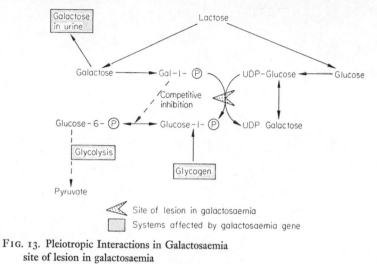

FIG. 13. Pleiotropic Interactions in Galactosaemia
site of lesion in galactosaemia
systems affected by galactosaemia gene

indeed to describe the process in great detail since so many books can be consulted, such as that by Watson [149] or Sargent [114]. Fig. 14 is a diagram representing the major steps in the synthesis of a protein involving two different types of polypeptide chain, as occur in so many proteins such as haemoglobin or lactate dehydrogenase. The hypothetical protein considered here also has prosthetic groups, which might be for example haem or flavin groups. In the complex series of reactions which are involved controlling influences could act at a number of points. The first stage to consider is **transcription** in which the DNA strand in the gene responsible for the production of the protein gives rise to an RNA molecule with a base sequence complementary to that in the transcribed strand of the DNA. This RNA molecule may eventually pass from the nucleus into the cytoplasm and there act as a template for the specification of the amino acid sequence in the protein. However, the RNA which is synthesized on the DNA template may be modified before passing into the cytoplasm. For example large RNA molecules may be synthesized and broken down to smaller molecules before leaving the nucleus, and there is evidence that a considerable amount of RNA is in fact totally hydrolysed without ever leaving the nucleus. Much of the RNA synthesized covers a wide range of molecular weight and this class of RNA is known as **nuclear heterogeneous RNA**.

It has been suggested by Scherrer and his colleagues [117] that the large molecules of the nuclear heterogeneous RNA may be broken down in stages until the functional messenger RNA molecules are formed and that there may be significant control mechanisms that operate during this processing of the nuclear RNA.

It appears that transcription, the initial stage in the sequence of macromolecular syntheses between the gene and the completed protein, represents a very significant stage in the regulation of cytodifferentiation. Much of the evidence available suggests that in cells which have differentiated along divergent pathways, different genes may be transcribed, so that dissimilar populations of messenger RNA molecules are found in the cytoplasm. It has occasionally been suggested that all of the genes are transcribed in the nucleus and that there is a selective breakdown of RNA molecules so that only some messenger molecules reach the cytoplasm. At present there is no compelling evidence for this and it would appear to be more economical of energy for only those

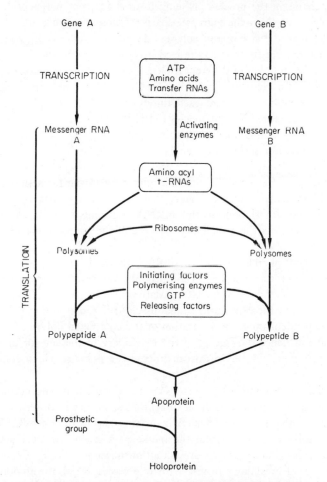

FIG. 14. Information Flow in the Synthesis of a Typical Protein

genes which are expressed to be transcribed in the first place. As we shall see, the weight of experimental evidence now favours the view that only selected genes are transcribed in differential cells.

Once the messenger RNA (mRNA) molecules have entered the cytoplasm they are involved in a complex series of reactions which ultimately give rise to a polypeptide chain with an amino acid sequence specified by the nucleotide sequence in the mRNA.

The mRNA molecules combine with ribosomes, which apparently pass along the length of the mRNA molecule as they are involved in the synthesis of the polypeptide chain. The bases in the mRNA interact by complementary base pairing with the transfer RNA molecules (tRNA), each of which is specifically charged with a single amino acid. A number of proteins associated with the ribosome catalyse the polymerization of the amino acids to form peptide bonds. Several ribosomes may be attached to the mRNA molecule at any one time, the number depending apparently upon the length of the mRNA molecule. Such aggregates of ribosomes, mRNA and polypeptide chains in the process of being formed (nascent polypeptide chains) are known as polysomes. Since the aggregate size is a function of the length of the messenger, it is related to the size of the polypeptide being synthesized. Large molecules, such as myosin or collagen, are synthesized by large polysomes.

A considerable number of different molecules and cellular structures are involved in the formation of an active polysome: about twenty different amino acids may constitute the polypeptide chain and each different amino acid will require the participation of one or more different species of tRNA molecule. Moreover, each type of tRNA has a corresponding enzyme capable of catalysing its attachment to the amino acid. The condensation of the amino acids into peptide bonds is catalysed by further enzymes. Now it would be possible to prevent the synthesis of proteins by withholding any of the score or so of molecules involved in the formation of the active polysomes. As will be discussed later, there is evidence that in some systems these steps in the translation of the mRNA molecule to form protein may be regulated by controlling the availability of some of the participant molecules in this process.

The attachment of ribosomes to the mRNA molecule at the very beginning of a cycle of protein synthesis involves certain proteins known as **initiation factors,** and these represent another point at which translation may be controlled, while at the end of the synthetic cycle the continuing availability of ribosomes and mRNA for protein synthesis is dependent upon the separation of the completed polypeptide chain from the polysome and hence the freeing of the machinery of the cell for more synthetic reactions. Once more, the **termination** process is a point at which control may be exercised.

In bacterial cells the mRNA molecules are usually unstable and become degraded within a short time of their utilization in synthesis, so that in order to maintain the rate of protein synthesis it is necessary to maintain a supply of new mRNA molecules. In higher organisms it appears that the messengers are much more stable than in bacteria, but it also seems that they are not all of the same stability. Clearly the regulation of the rate of breakdown of mRNA is one way in which the amount of protein in a cell can be controlled and we shall see later some examples of the way in which such regulation may act during differentiation, for instance in the lens fibre cell.

A final point in the scheme in Fig. 14 which can in some circumstances exercise a regulating influence is the attachment of any prosthetic groups which convert the **apoprotein,** composed purely of polypeptide chains, into a biologically active protein molecule (**holoprotein**). Not all enzymes have such prosthetic groups, but there are some clear instances where the availability of such groups exerts a major influence on the rate of synthesis of the molecule, for instance the effect of the presence of haem on the synthesis of the globin chains which are utilized in the formation of haemoglobin (see Chapter 6).

As more different systems of protein synthesis are studied in detail it becomes apparent that the different stages in translation of the RNA message into protein are themselves regulated in fairly precise ways. There is much evidence that the mere presence of a specific messenger in a differentiating cell is not in itself sufficient for the protein for which it codes to be synthesized. In a number of quite different cells it seems that the messenger may be present well before it is used and that it is the absence of other components, or maybe the presence of specific inhibiting agents, which prevent the synthesis of the protein. Some of these examples will be discussed later. As has been mentioned, a further aspect of the translational control of protein synthesis is concerned with the stability of the mRNA. Even so, it is most probable that the fundamental process which determines the pattern of protein synthesis within a cell is the regulation of the transcription of DNA, and that the translational control mechanisms are chiefly concerned with the timing of specific protein synthesis. The evidence for the significance of the differential transcription of genes will now be considered.

Changes in the RNA content of cells during differentiation

Our knowledge of the changing composition of the RNA of cells is very much limited by the techniques that are available for investigating RNA. It is useful to consider briefly what these methods are, since we can then assess better the experiments that have been carried out with differentiating cells. The RNA molecules in cells differ in size, in location within the cell, in the relative proportions of the different bases and in the sequence of these bases along the polynucleotide chain, and in their function in the metabolism of the cell. The bulk of the RNA in the cytoplasm of the cell is that associated with the structure of the ribosomes and this is of various molecular weights, while another class of cytoplasmic RNA is that containing all the different molecular species of transfer RNA. The different size classes of RNA can be fairly readily separated, for example by centrifugation on a sucrose density gradient, by chromatography on columns of methylated albumin adsorbed to kieselguhr (MAK columns), or by electrophoresis on polyacrylamide gels. A diagram of the results of such a separation is seen in Fig. 15 [19]. The ribosomal RNA (rRNA) forms fairly distinct groupings when using these separation techniques, while the different transfer RNA species, even though they differ in base sequence and function, also form a distinct group. Thus we can fairly easily study changes in these classes of RNA during development, without confusion arising among the different classes. Much work has been done on the RNA content of embryos at different stages of development and the results are summarized in a review by Gurdon [52]. In many species there is little synthesis of rRNA during early development. For instance, in the toad *Xenopus laevis* rRNA synthesis does not begin until gastrulation, while tRNA synthesis begins at the blastula stage.

The fractionation of the different tRNA molecules presents rather greater technical problems. Pioneering work was done by Holley, using two-phase partition methods, while recently new chromatographic systems have been developed to separate different species of tRNA. By these methods it has been established that the relative proportions of the tRNA molecules specific for the different amino acids may

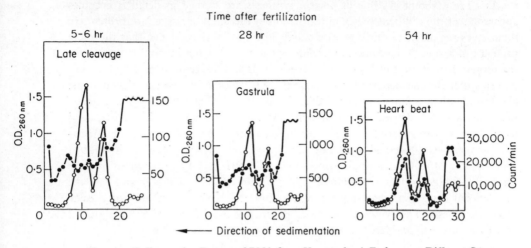

FIG. 15. Sedimentation Pattern of RNA from *Xenopus laevis* Embryos at Different Stages
Each sucrose density gradient centrifugation was performed on RNA from 150 embryos.
optical density at 260 nm; radioactivity
From Brown, D.D. and Littna, E.: RNA synthesis during the development of *Xenopus laevis*, the South African clawed toad. *Journal of Molecular Biology* 8, 669–87 (1964).

change during development [131] (Fig. 16). This provides support for the suggestion that protein synthesis may be regulated by controlling the availability of tRNA species [65]. The role of tRNA in differentiation has been reviewed by Sueoka and Kano-Sueoka [130].

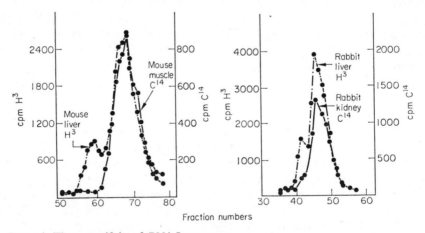

FIG. 16. Tissue-specificity of tRNA Isoacceptors
Chromatographic patterns obtained from serine-tRNA from mouse liver and muscle and rabbit liver and kidney.
From Taylor, M.W., Granger, G.A., Buck, C.A. and Holland, J.J.: Similarities and differences among specific tRNA's in mammalian tissues. *Proceedings of the National Academy of Sciences of the United States of America* **57**, 1712–19 (1967).

34

These changes in the pattern of synthesis and cellular content of rRNA and tRNA during development are probably significant in controlling the overall rates of protein synthesis, but it is likely that they are insufficiently specific to discriminate in their effects among different proteins with any great subtlety, and so they are probably not responsible for the difference in the patterns of protein synthesis among differentiated tissues. The mRNA is a more likely candidate for the agent that is capable of such specific regulation of protein synthesis.

The fractionation and assay of mRNA presents many problems. In most types of cells the total amount of mRNA present is probably only a small fraction of the rRNA and the tRNA of the cell. Moreover, the mRNA molecules for different proteins would be expected to differ in both size and base sequence according to the protein that they synthesize. Thus the separation techniques that operate by size and which gave such good results with rRNA and tRNA are generally less effective in characterizing mRNA, especially as some of the size classes of messenger may overlap with some of the sizes of rRNA. While the total amount of mRNA in the cell may be small, the mRNA for a particular protein will generally be an even smaller proportion of the total RNA. Added to these difficulties is the fact that mRNA appears to be particularly easily degraded during its isolation from the cell.

Since size cannot be used as a criterion of mRNA in most cells, attempts have been made to estimate changes in the messenger content of cells by a method which is dependent upon the base sequence of the RNA. The best method at present available for making such discrimination is the technique of **nucleic acid hybridization** (Fig. 17). We know from the classical study of DNA structure of Watson and Crick that the four principal bases of DNA are capable of forming hydrogen bonds between each other in specific pairs while they are combined in the polynucleotide chains, adenine binding to thymine and guanine to cytosine. Thus DNA molecules are in their most stable state when the bases are paired in this way. Similar base pairing by hydrogen bond formation can occur between adenine and uracil, the analogue of thymine, and such binding of adenine to uracil bases occurs naturally during the course of RNA synthesis on a DNA template. When the hydrogen bonds which hold the bases in these pairs are broken (**denaturation**), either by raising the temperature or by treatment with agents such as formamide, then the two complementary strands of the DNA can be separated, while lowering the temperature or removing the denaturing agent may under suitable conditions permit the strands to associate together again (**renaturation**). Provided the conditions are carefully controlled, it is possible for the corresponding bases to align themselves in a quite specific manner. If RNA molecules are added to DNA solutions which have been dissociated into their separate polynucleotide strands, then specific base pairing can occur between the RNA and complementary portions of DNA, and hybrid molecules containing one strand derived from DNA and one derived from RNA can be formed. This phenomenon can be used as a method for distinguishing the extent to which corresponding base sequences exist in RNA and DNA molecules.

The usual method by which the binding of RNA to DNA is measured is to make one of the molecules radioactive. If the RNA is derived from cells which have been grown in a medium containing radioactive phosphate, $^{32}PO_4$, then the binding of radioactivity to the DNA can be taken as a measure of the extent of formation of

DNA-RNA hybrid molecules. If during the process of hybridization of radioactive RNA to DNA, non-radioactive RNA of similar base sequence is added to the system, then this would compete successfully with the radioactive molecules for the complementary sites on the DNA molecules, since the formation of specific base pairs will not discriminate between the radioactive and the normal molecules. If only a

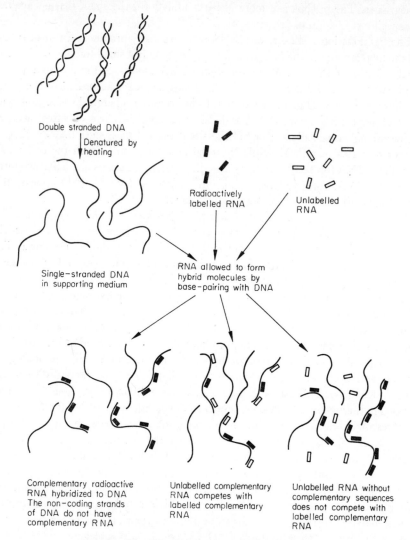

Double stranded DNA

Denatured by
heating

Radioactively
labelled RNA

Unlabelled
RNA

Single-stranded DNA
in supporting medium

RNA allowed to form
hybrid molecules by
base-pairing with DNA

Complementary radioactive
RNA hybridized to DNA
The non-coding strands
of DNA do not have
complementary RNA

Unlabelled complementary
RNA competes with
labelled complementary
RNA

Unlabelled RNA without
complementary sequences
does not compete with
labelled complementary
RNA

FIG. 17. The Principle of Competitive Hybridization

limited amount of DNA is available, then the addition of increasing amounts of unlabelled RNA will inevitably diminish the binding of radioactive RNA with the same sequence, whereas the addition of RNA with a quite unrelated sequence would not compete for the binding sites. Thus the extent of the competition between unlabelled RNA and labelled complementary RNA for binding sites on DNA molecules can be used as a measure of the similarity of base sequence between the RNA molecules. One of the earliest uses of this technique to test for differences in the RNA's of different tissues was that carried out by McCarthy and Hoyer in 1964 [83].

The composition of the RNA of different tissues

In one of their experiments McCarthy and Hoyer prepared RNA from a line of mouse cells in tissue culture (L-cells) which had been grown in the presence of radioactive phosphate. This RNA was allowed to form hybrid molecules with denatured DNA derived from mouse tissue. If unlabelled RNA derived from L-cells grown in non-radioactive medium was added to the system during hybridization there was a competitive inhibition of the binding of radioactivity to the DNA, and as can be seen from Fig. 18, increasing amounts of RNA from L-cells reduced the binding of radioactivity considerably. When RNA from either liver or spleen was added to the hybridizing system, then the inhibition of binding was less than that produced by RNA from L-cells, and the competitive effect of liver differed from that of spleen (Fig. 18). This experiment, and others by the same workers, showed that the RNA content of different tissues varies in its capacity to hybridize with DNA and so probably differs in the base sequences that are present.

Hybridization of RNA in developmental studies

The way in which RNA changes in an embryo during development has also been studied by hybridization competition methods similar to those of McCarthy and Hoyer. Denis [33] investigated changes in the embryos of the toad *Xenopus laevis*. DNA from this toad was prepared, and RNA was used which was isolated from tadpoles of a certain stage (stage 42 in the normal tables of embryonic development) which had been labelled by growing the tadpoles in medium containing bicarbonate labelled with ^{14}C. The hybridization was carried out between this RNA and the DNA, and various amounts of unlabelled RNA were used as competitors which were derived from earlier embryonic stages. Fig. 19 shows that the degree of competition offered by these RNA preparations alters during development. The conclusion that Denis drew from these experiments was that the messenger RNA present in unfertilized or early cleavage eggs had no detectable nucleotide sequences in common with the mRNA of later embryos, and that during development there is an increasing amount of RNA which has sequences similar to those of fully differentiated organisms. From other experiments using the same technique Denis concluded that some of the RNA which is present in the gastrula stage is absent from later stages. Thus the development and differentiation of the embryo is accompanied by changes in the composition of the RNA, including both the loss of some types of RNA and the acquisition of new types.

Limitations of the hybridization technique

Although it was originally hoped that the technique of RNA-DNA hybridization would provide a measure of changes in the population of mRNA in cells, it is now realized that major difficulties stand in the way of doing this. There are

37

evidently portions of the DNA in the genome of higher organisms which have base sequences that are highly repetitive, the same sequence of bases being represented perhaps as many as 100,000 times [18]. These regions of DNA, which may constitute 5–10 per cent of the genome, have a higher affinity for their complementary RNA than do those regions of the genome with a greater variety of base sequences. During nucleic acid hybridization the repetitive sequences combine very rapidly, and under

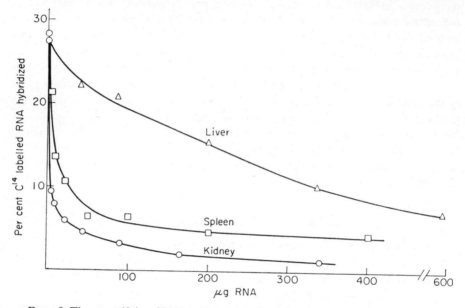

FIG. 18. Tissue-specificity of RNA as Revealed by RNA-DNA Hybridization
    Competition by unlabelled RNA in the reaction of purified ^{14}C-labelled RNA from a primary culture of kidney cells with DNA.
    From McCarthy, B.J. and Hoyer, B.H.: Identity of DNA and diversity of messenger RNA molecules in normal mouse tissues. *Proceedings of the National Academy of Sciences of the United States of America* **52**, 915–22 (1964).

many circumstances the RNA associated with the repetitious DNA may be the only RNA which hybridizes. The RNA which directly specifies protein structure will be among that with the most varied sequences and so will be relatively difficult to hybridize.

    In view of this heterogeneity of the rate of hybrid formation we can draw only limited conclusions from the type of experiment done by McCarthy and Hoyer and by Denis. An attempt to use hybridization methods more specifically to measure changes in the synthesis of non-repetitive RNA was made by Grouse, Chilton and McCarthy [49]. The technique that they used involved purified unique sequences of DNA hybridized with excess RNA. This method is believed to avoid the problem of preferential hybridization of repetitive sequences. These workers found that the RNA of various tissues differed to a certain extent, but that some of the RNA sequences were common to the various organs tested. In the case of brain, changes in RNA composition during development were detected, but liver RNA showed little change with age.

It is evident from a variety of experiments that there is a change in the types of RNA molecules present in the tissues during development, but there is not much direct experimental support for the belief that differentiation involves the formation of new populations of mRNA molecules in the cell. There is a great body of indirect evidence from the wide field of molecular biology which leads us to believe

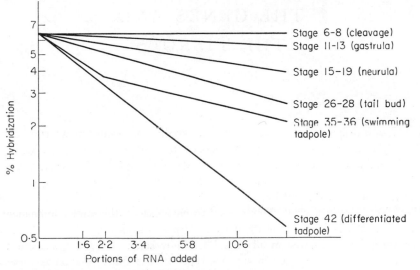

FIG. 19. Changing Composition of RNA during Development
Competition experiment between pulse-labelled RNA from stage 42 tadpoles and non-labelled RNA from earlier embryos.
From Denis, H.: Role of messenger ribonucleic acid in embryonic development. *Advances in Morphogenesis* **7**, 115–50 (1968). Copyright (1968) Academic Press, New York.

that as cells differentiate they will contain changing populations of mRNA but the absence of much compelling and direct evidence of this is one of the major gaps in our present knowledge of developmental biology. The significance of the changing population of RNA derived from the repetitious DNA will be discussed in Chapter 7.

Summary

The control of the metabolism of a cell is the result of complex interactions of many factors. Certain key enzymes catalysing rate-limiting steps in metabolic pathways may be particularly important. The activity of such enzymes is influenced by several agencies, but during differentiation the control of the rate of synthesis of the enzyme is often a major determinant. The synthesis of enzymes and other proteins may be regulated at several levels in the flow of information from gene to protein, but in many systems it appears that the control of the synthesis of messenger RNA is the main directing factor in differentiation. The control of the rate of translation of this message into protein is also significant in many systems.

4

THE GENES AND
THE TISSUES

There is a widespread belief that when nuclear division occurs in a fertilized egg, ultimately giving rise to a multicellular embryo, the **mitotic divisions** of the nuclei divide the chromosomes exactly between the daughter nuclei. The result is a mass of cells which are all genetically identical with the fertilized egg from which they were derived. By this mitotic nuclear division the differentiated cells in the body of a higher plant or animal would be expected to have exactly the same complement of genes. Since the various cells of the body can be seen to differ, it follows that if the genes are indeed the same in all cells, then only some of the genes of the total gene population are expressed in any one type of cell; differentiation is therefore a matter of the varied expression of the genes in different types of cell. This chapter examines some of the evidence both for and against this view that the **genome**, or total complement of genes, is constant throughout the body of an organism.

The nature of mitosis

Cytological evidence indicates that, in the great majority of cases, mitotic nuclear division is a mechanism whereby the chromosomes of daughter cells are replicated so that they are identical to those of the parent cell. In this respect mitosis contrasts with **meiosis** during which the chromosome number of the nuclei is halved and also rearrangement of genetic material within the chromosomes may occur by the process of crossing-over. In the absence of the comparatively rare event of mutation, it is believed that daughter cells produced by mitosis are genetically identical with their parent cell, and that this constancy of the genotype is maintained even after many cycles of mitosis have taken place to produce a line of cells. Such a mass of cells produced by mitosis from a single parent cell and all having the same genetic constitution, or genotype, constitutes a **clone**.

The maintenance of the genotype through mitotic cell divisions is exemplified by those organisms which are capable of asexual reproductive processes which involve neither meiosis nor fertilization but only mitotic cell divisions. Such organisms resemble their parent in genotype and so are usually very similar in their observable characteristics, or **phenotype**.

If mitosis propagates the genome without modification, then all of the diploid nuclei of an individual should contain the same amount of DNA, while the haploid cells of the gametes should contain half this amount. The average DNA content of nuclei has been calculated by measuring the total DNA in a suspension of a known number of nuclei isolated from different tissues of the ox. Some results are shown in Table 6.

TABLE 6. DNA Content of the Nuclei of Various Bovine Tissues

	Wt. of DNA/nucleus (g. $\times$ 10^{-12})
Thymus	6·6
Liver	6·4
Pancreas	6·9
Kidney	5·9
Sperm	3·3

Based on data of Vendrely [146]

It was considered that the variation among the diploid tissues was within experimental error and concluded that the DNA content of nuclei within a species was constant except in the haploid cells such as the sperm, and that in these haploid cells the DNA content was half that of the diploid cells. Further results with other species confirmed this impression, and also indicated that the nuclei of different species showed marked variation in their DNA content. This is discussed in Chapter 7. These experiments have been summarized in a review by Vendrely [146].

Chromosome constancy

The function of mitosis in maintaining an identical pattern of genetic material in a line of cells, even though the cells may be differentiated cytologically, is further confirmed by a study of the **karyotypes** (the number and morphology of the chromosomes) of different tissues of the body. For example Tjio and Puck examined eight different tissues in man and the chromosomes were counted in the nuclei of 1,825 cells. In one nucleus 45 chromosomes were seen and in another there were 47, while in some of the nuclei **tetraploidy** was detected, 92 chromosomes being present. Nevertheless, in the vast majority of cases 46 chromosomes were found, and moreover the microscopic appearance of the chromosomes did not differ from tissue to tissue [136].

Banding pattern of polytene chromosomes

In some tissues of the larvae of certain insects, chiefly some species of flies (Diptera), the chromosomes are present in an unusual state, there being a great increase in the number of strands compared with the normal diploid nucleus. These chromosomes, which are described as **polytene**, are in a very extended state and their DNA content may be about one thousand times that of the normal adult chromosome. Such a variation in DNA content and chromosome structure is, of course, a contradiction of the view that the chromosomes remain constant in different tissues and at different stages of development. Nevertheless, in a paradoxical way these chromosomes provide evidence that the genes which are present in one tissue are to be found similarly arranged in other tissues. Each polytene chromosome has transverse **bands** which stain deeply with some histological dyes, and these bands are separated by areas staining more faintly. The bands vary in their spacing and each chromosome has its own characteristic pattern of bands (Fig. 20). Although the precise relationship

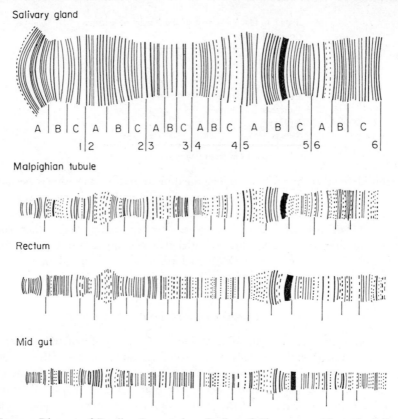

FIG. 20. Diagram of Banding Patterns in a Portion of Chromosome No. 3 from Four Tissues of *Chironomus tentans*

From Beermann, W.: Chromomerenkonstanz und spezifische Modifikationen der Chromosomenstruktur in der Entwicklung und Organdifferenzerung von *Chironomus tentans. Chromosoma* (Berl.) 5, 139–98 (1952). Berlin-Heidelberg-New York: Springer.

between the transverse bands and the genes on the chromosome is not clear, cytogenetic techniques have established that there is a relationship between changes in genetic maps, for instance in translocations and deletions, and changes in the banding pattern of polytene chromosomes. Beermann [7] in a very detailed study of the larval chromosomes of the midge *Chironomus tentans* has compared the banding patterns in chromosome number 3 in the salivary gland, Malpighian tubule, rectum and mid-gut. Although there is some variation in the transverse width of the chromosomes from one tissue to another, the general pattern of banding in the four tissues examined was found to be remarkably similar.

We shall return later to the structure and function of the polytene chromosomes during differentiation, but for the moment they are mentioned as further evidence for the view that the same genes are present in various tissues of the body.

Studies of DNA structure

If the genome of all the tissues of an organism is identical, then the composition and structure of the DNA of the nuclei of one tissue should be identical with that of other tissues. This is not an easy matter to demonstrate, and it is perhaps wisest to say that no differences among the DNA's of different tissues have been detected, except in certain cases to be discussed later.

The sensitive technique of hybridization of DNA molecules has been employed by McCarthy and Hoyer [83]. This method, which is analogous to the hybridization with RNA described in the previous chapter, depends on the capacity of separated strands of DNA to recombine in correct register by the formation of hydrogen bonds and on the fact that the greater the degree of complementarity between the separated strands, the more fully this recombination can occur. Strands of DNA derived from the same or identical molecules will be fully complementary, each adenine pairing with a thymine in the other strand and each cytosine pairing with a guanine. DNA strands derived from different molecules will not match up so readily and so will not hybridize to the same extent. In practice even the hybridization of fully complementary molecules is not totally efficient and allowance must be made for this.

McCarthy and Hoyer [83] prepared DNA from mouse embryos, denatured it so as to separate the strands, and then trapped these strands in an agar gel. Radioactively labelled DNA prepared from mouse embryos and from mouse cells maintained in tissue culture was mixed with the DNA trapped in the agar under conditions which favoured the interaction of DNA strands by the formation of hydrogen bonds between complementary bases. Under these conditions some 15 per cent of the labelled DNA formed hybrid molecules with the unlabelled DNA trapped in the agar. If unlabelled DNA identical (**homologous**) with the radioactive DNA was added to the mixture before hybridization occurred, it competed successfully with the labelled DNA for the complementary sites upon the DNA trapped in the agar: in contrast, the addition of unlabelled DNA without any sequence homologies (**heterologous**) with the labelled DNA did not have any effect upon the hybridization. McCarthy and Hoyer demonstrated that while DNA from the bacterium *Bacillus subtilis* had no effect on the hybridization of the molecules of mouse DNA, unlabelled DNA isolated from seven different tissues of the mouse competed successfully with the radioactive mouse DNA for hybridization sites on the DNA strands trapped in the agar. Moreover, the extent of the competition was the same for the DNA from the various tissues tested (Fig. 21). The authors concluded, therefore, that no differences could be detected among the base sequences in any of the mouse DNA's tested.

The totipotency of individual cells

The diploid zygote formed at fertilization is a single cell which is capable of division and differentiation to give rise to the total range of cell types found in the adult organism. It is therefore said to be **totipotent**, having the potentiality of forming all the types of cells in the body. If the entire genome is conserved during mitosis, then under suitable conditions it might be possible to stimulate an already differentiated cell to develop in the same way that the zygote does. On the other hand, if during cytodifferentiation there is a loss of genetic potential and the nuclei themselves become differentiated in the sense of losing the capacity for directing the synthesis of some proteins, then a differentiated cell would be incapable of ever giving rise to cells which would in turn form a complete organism. It might be expected that it would be difficult to find the correct conditions for such a demonstration of the totipotency of a single cell, and a negative experiment could not be regarded as conclusive. Even

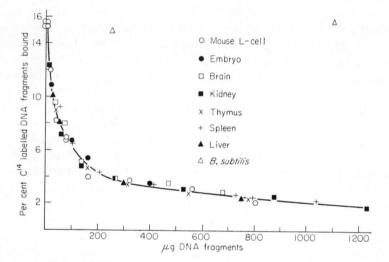

FIG. 21. Similarity of DNA from Different Mouse Tissues as Revealed by DNA-DNA Hybridization

The inhibition of hybridization between mouse DNA molecules caused by the addition of DNA from different tissues.

From McCarthy, B.J. and Hoyer, B.H.: Identity of DNA and diversity of messenger RNA molecules in normal mouse tissues. *Proceedings of the National Academy of Sciences of the United States of America* **52**, 915–22 (1964).

so, in some cases it has been possible to achieve the formation of a complete organism from a single differentiated cell, thereby giving the most conclusive proof that the total genome remains present even after differentiation. So far all the cases of regeneration of an entire organism from single cells have been in plants: no comparable demonstration is available with animal material as yet.

The first case studied was the crown-gall tumour of the tobacco plant in which it was found that clones of cells derived from a single tumour cell could, in some circumstances, give rise to roots and leaves [15]. A more extensive study, in which cells derived from non-tumorous differentiated tissue have been used, is to be found in the work of F.C.Steward and his colleagues who cultured small pieces of tissue from the secondary phloem tissue of the carrot, *Daucus carota* [129]. Single cells were obtained from the cultured mass of tissue and were then cultured in a synthetic medium supplemented with coconut milk, which contains a number of substances which stimulate cell division. In this medium the single cells divided to produce a clone of cells whose structure resembled the early embryo of a normal carrot. This embryo-like mass of cells passed through a series of changes as the cells continued to divide, so that the various forms of the normal embryonic development were represented. After some time the cell mass produced roots, and then it was transferred to a solid medium on which its growth continued, and a shoot was formed. Ultimately a complete carrot plant was formed with storage organ, leaves and flowers, and seeds were produced. No more complete demonstration could be desired to show that a single differentiated plant cell retains the full genetic potentiality of the zygote (Fig. 22).

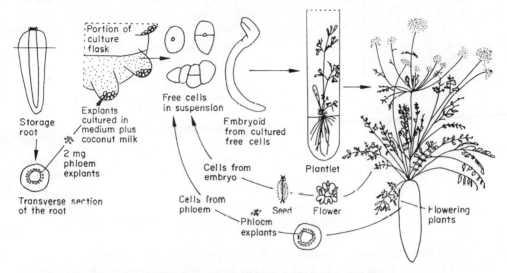

FIG. 22. Totipotency of Differentiated Plant Cells Demonstrated in Cell Culture
   The origin of a complete differentiated plant from cultured cells of the phloem of the carrot.
   From Steward, F.C., Mapes, M.O., Kent, A.E. and Holsten, R.D.: Growth and development of cultured plant cells. *Science* **143**, 20–7 (1964). Copyright (1964) by the American Association for the Advancement of Science.

Since Steward's initial work with the carrot, the production of complete plants from the culture of single differentiated cells has been achieved using cells of other parts of the carrot, with cells of the water parsnip, *Sium saure* [128] and the tobacco plant, *Nicotiana* [145]. These experiments provide clear evidence of the totipotency of the nuclei of differentiated plant cells. In the case of animals the evidence available is much less direct.

Regeneration of animals

It is well known that parts of plants when removed can regenerate and grow into whole plants, and we saw above that even a single cell may have this potentiality. With the lower animals powers of regeneration are also remarkable, but we must beware of drawing excessive conclusions from these observations because it frequently happens that regeneration may be from undifferentiated cells which have been retained mingled with the differentiated cells of the body. Thus many instances of regeneration in animals do not provide clear evidence of the potentialities of differentiated cells. The capacity of a planarian worm to regenerate half a body after it has been cut in half is largely a result of the division and differentiation of cells which were previously undifferentiated. The cases of regeneration of limbs and tail in newts may also involve undifferentiated cells. Even so there are in the animal kingdom cases where it is well established that the regeneration of an organ composed of one type of cell is achieved from cells which are fully differentiated in another direction. Some of the best examples of this regeneration by dedifferentiation and ensuing redifferentiation are to be found among the tissues of the eye of amphibians.

In the newt *Triturus pyrrhogaster* surgical removal of the lens of the eye is followed by a regenerative process in which a new lens is formed from cells of the iris of the eye. The changes involved show clearly that differentiated iris cells are involved in the lens formation (Fig. 23); the cells of the margin of the iris near the pupil of the eye undergo ultrastructural changes including a change in the shape of the nucleus, and enlargement of the nucleolus. The melanin pigment granules which are characteristic of the iris cells are either transferred to amoeboid cells within the iris or they are discharged directly into the cavity of the eye from the iris epithelial cells [36]. After depigmentation the iris cells divide by mitosis and form a hollow ball of cells, similar to the vesicle formed during the normal embryonic development of the lens from the embryonic ectoderm. Cells of this vesicle then differentiate by elongation to form fibres, which contain the protein γ-crystallin which is a characteristic lens protein [37]. A detailed account of the differentiation of the lens fibre cell will be found in Chapter 6.

In other species of amphibia, different eye tissues are involved in lens regeneration. For example, in the toad *Xenopus laevis* either iris, retina or cornea cells can give rise to lens cells after dedifferentiation. In all of these cases of lens regeneration the lens cells are formed from cells which had previously been fully differentiated in the iris, retina or cornea, and though such cells do not normally synthesize lens proteins in quantity, they have nevertheless retained the genes necessary for such syntheses and in certain circumstances these genes can still be expressed.

The totipotency of early cleavage nuclei

It has been known for some time that nuclei retain their totipotency at least for some small number of cell divisions after the formation of the zygote. The existence of identical twins in man shows that after one cell division both nuclei still have the full complement of genetic material of the zygote, and the results of some embryological experiments show that this totipotency can persist somewhat longer.

In the sea urchin embryo the cells formed by cleavage of the egg cytoplasm and division of the nuclei, before there is any major cell differentiation, are known as **blastomeres**. At the two-cell stage the two blastomeres can be separated and each capable of developing into a complete embryo. This capacity for complete develop-

46

ment is retained at the four-cell stage; even at the eight-cell stage it has been possible to isolate a blastomere and find development of a pluteus larva that although smaller than normal, is nevertheless complete in all its features (Fig. 24) [63].

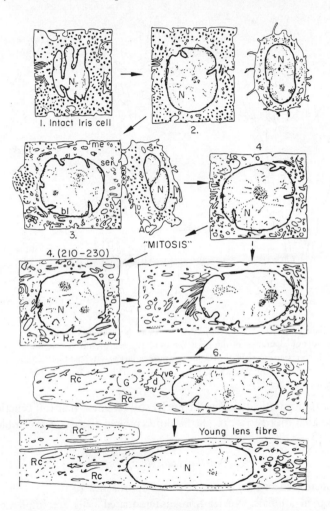

FIG. 23. Dedifferentiation of an Iris Cell and its Redifferentiation as a Lens Fibre in Lens Regeneration in the Newt
    From Eguchi, G.: Electron microscopic studies on lens regeneration II. Formation and growth of lens vesicle and differentiation of lens fibres. *Embryologia* (Tokyo) 8, 247–87 (1964)

The totipotency of amphibian nuclei, at the sixteen-cell stage, was demonstrated in an ingenious experiment by Spemann [124]. Using fertilized eggs of the newt *Triturus*, he tied a fine hair around the cleaving egg so as to restrict the nucleus to one part of the cytoplasm. Nuclear division proceeded normally and there was cytoplasmic cleavage in the nucleated part of the egg, until sixteen blastomeres were formed on that side of the ligature. During these early nuclear divisions the size of the nucleus became somewhat reduced and at the sixteen-cell stage a nucleus was able to pass through the ligature into the region of cytoplasm previously devoid of nuclei. Spemann then tightened the hair loop sufficiently to separate the two portions of the egg. In a number of cases complete embryos developed from both portions of the egg, indicat-

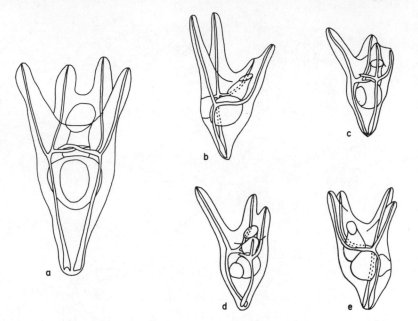

FIG. 24. Totipotency of Sea Urchin Blastomeres at the Four-Cell Stage
    The effects of separation of the blastomeres of the sea urchin egg at the four-cell stage (a) normal larva; (b to e) smaller but otherwise normal larvae obtained from single blastomeres.
    After Hörstadius and Wolsky [63] from Balinsky, B.I.: *An Introduction to Embryology*. Philadelphia: Saunders.

ing that after four mitotic divisions the nuclei still retained their full genetic potentiality and were able to give rise to all the differentiated tissues of a complete embryo (Fig. 25).

Nuclear transplantation

This has constituted one of the most direct approaches to the problem of nuclear differentiation in animals. Nuclei from differentiated cells are implanted into egg cells, the original nuclei of which have been either removed or inactivated by irradiation. The subsequent development of such egg cells then provides an indication of the potentiality of the implanted nucleus.

The experiments that most clearly indicate that the nucleus does not undergo any irreversible differentiation or loss of genetic capability during cell differentiation are those carried out at Oxford by Gurdon and his group [50, 53]. In these experiments egg cells of the toad *Xenopus laevis* were irradiated to inactivate the nucleus. Using a micropipette, nuclei were removed from the cytoplasm of cells from the epithelium of the intestine of tadpoles. These cells donating nuclei appeared to be fully differentiated epithelial cells with the characteristic brush border. The nuclei, together with a very small amount of cytoplasm which was unavoidably taken up with the nucleus, were then injected into the irradiated eggs, which were then allowed to develop. Frequently these eggs failed to develop, but in some cases, rarely more than 35 per cent of those injected and sometimes as few as 1 per cent, normal embryonic development did take place. Control experiments in which a small piece of cytoplasm but no nucleus was injected confirmed that the nucleus was the significant component

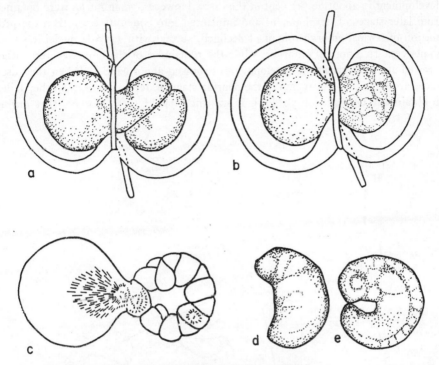

FIG. 25. Totipotency of Amphibian Nuclei at the Sixteen–Cell Stage
    Formation of an embyro of *Triturus* from a nucleus derived from the sixteen–cell stage
(a) The egg constricted by a hair loop, with cleavage beginning in the nucleated half
(b) Stage of penetration of a nucleus into the uncleaved half of the embryo
(c) Diagrammatic section of same stage as (b)
(d) Embryo formed from cytoplasm and nucleus from sixteen–cell stage
(e) Embryo formed from cytoplasm which contained zygote nucleus
    After Spemann, from Balinsky, B.I.: *An Introduction to Embryology*. Philadelphia: Saunders.

transplanted and that the small amount of cytoplasm involved could be disregarded. On some occasions the development of the eggs with transplanted nuclei was followed for a time sufficient to demonstrate that fertile male and female toads were produced (Fig. 26).

    The technical difficulties in this kind of experiment are very great and the nucleus is easily damaged while it is outside the cell. Gurdon attributes failure of development after nuclear transplantation to probable nuclear damage during handling, and hence we should perhaps concentrate our attentions on those instances in which there was successful development rather than on the more numerous occasions when development was abnormal. If we do this then we must conclude that the nuclei of apparently fully differentiated cells still retain their genetic totipotency.

A somewhat different conclusion has been drawn about nuclear differentiation by other workers using a similar technique. Briggs and King and their colleagues in Bloomington, Indiana, have studied nuclear transplantation in species of *Rana* for many years. If they took nuclei from the blastula stage of the frog *Rana pipiens* and implanted them into enucleated eggs of the same species they obtained normal development in about 80 per cent of the cases. However, when nuclei were obtained from later stages of development and implanted into enucleate eggs, then the proportion of normal embryos obtained declined, so that with gastrula nuclei less than 20 per cent gave normal embryos, while the remainder of the embryos were either arrested at an early stage of development or showed abnormalities. When they used nuclei from the mid-gut region of the neurula stage of development, then they failed to obtain any normal embryos [71]. It was the view of these workers that during

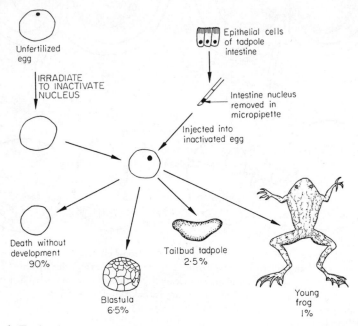

FIG. 26. Testing for Nuclear Differentiation by Nuclear Transplantation in Amphibia
   The figures represent the proportions of eggs with transplanted nuclei which survive to the particular stages.

cell differentiation changes occurred in the nucleus, including modifications of the chromosomes, such that the nucleus lost its totipotency and was no longer able to express all of the genetic potentialities that were to be found in the nucleus of the zygote.

In a more recent detailed study of the chromosomes of nuclei from differentiated cells Di Berardino and Hoffner [34] have come to the conclusion that by the time of gastrulation in *Rana pipiens* there have been changes in the nuclei of endodermal cells which prevent normal nuclear replication and division if the nuclei of these cells are transplanted into egg cytoplasm. These chromosomal changes involve the elimination of variable amounts of genetic material so that the genome of individual differentiated cells is less complete than that of the zygote or the cells formed by the first few cleavage divisions.

It is not easy to resolve the conflicting evidence of the various experiments on nuclear transplantation. We may concede that many factors could prevent a successful nuclear transplantation and suggest that in the more differentiated cells it becomes more difficult to achieve the effective removal and implantation of nuclei (see Table 7).

TABLE 7. Success of Nuclear Transplantation with Nuclei from Different Sources

Donor nuclei	No. of transfers	Percentage survival to		
		Blastula	Tail-bud tadpole	Young frog
Rana pipiens				
Late blastula	82	55	44	32
Early gastrula	151	62	32	26
Late gastrula	155	51	35	—
Tail-bud endoderm	130	7	2	—
Xenopus laevis				
Late blastula and late gastrula	279	62	48	35
Tail-bud endoderm	174	27		8
Intestinal epithelium	726	6·5	2·5	1

We would then place the emphasis on the few successes rather than the many failures and conclude that even in differentiated cells the nuclei retain the same complement of genes to be found in the zygote. This would then accord with the experimental demonstrations of the totipotency of differentiated plant cell nuclei and fall in with the suppositions built upon observations of the conservative effects of mitotic cell division and on the constancy of content, composition and hybridizing capacity of nuclear DNA.

EVIDENCE OF INCONSTANCY IN THE GENETIC MATERIAL DURING DIFFERENTIATION

Elimination of chromosomal material in Ascaris

Recent evidence, such as that of King and Briggs described above, that nuclei may become irreversibly modified during cellular differentiation corresponds with some observations made by microscopy by Boveri as long ago as 1887. Boveri showed that in the round worm *Ascaris* two groups of cells could be distinguished quite early in development: one group, the **germ line**, eventually gives rise to the cells which produce the gametes, while the other group of cells, the **somatic** cells, gradually become differentiated in various ways to make up the remaining tissues of the body. During nuclear division the nuclei which eventually form the germ line receive the full chromosome complement which is present in the zygote, whereas the chromosomes of the nuclei which are to form the somatic cells become modified, the ends of the chromosomes being eliminated during mitosis (Fig. 27). It therefore appears that in this particular animal the full chromosome complement is not necessary for the development of the somatic cells. Thus we have in *Ascaris* a case of visible nuclear differentiation. However, once the somatic cell nuclei have received their reduced complement of chromosomal material, there is no visible evidence of any further variation in the karyotype in the different types of somatic cell in the body [13].

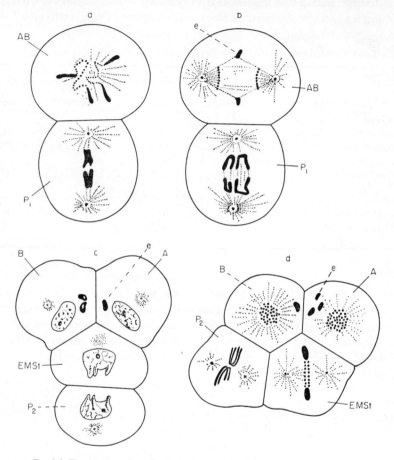

FIG. 27. Partial Chromosome Loss in Somatic Cells of *Ascaris*
(a) beginning of the second cleavage
(b) later stage of the same cleavage, with the chromosomes becoming fragmented, and their ends, *e*, lost in the cytoplasm
(c) Four-cell stage, viewed from animal pole
(d) Later stage of four-cell stage; chromosome fragmentation and loss of ends is occurring in cell *EMSt*.
     After Boveri [13], from Waddington, C.H.: *Principles of Embryology*. London: Allen and Unwin.

Other cases of elimination of chromosomal material

The fly *Sciara* gives another example of variation in the chromosome content of different tissues. In addition to the usual autosomes and to the sex chromosomes there are 'limited' chromosomes restricted to the cells of the germ line. During the first five or six cleavage divisions all of the chromosomes are represented in all of the cells, but after then the nuclei of cells destined to become somatic cells eliminate the limited chromosomes, while the cells going on to form the germ line maintain the complete set of chromosomes. The limited chromosomes are microscopically distinguishable from the remainder of the chromosomes, being larger both in diameter and length [86].

In the plant *Sorghum* there are anomalous extra chromosomes, known as *B* chromosomes, which do not pair at meiosis. During the cell divisions which give rise to the embryonic plant these *B* chromosomes become eliminated from the line of cells which go to form the root tissues, but they are present in some tissues of the shoot portion of the plant, including those cells which give rise to the flower tissue [68].

It should be emphasized that those cases outlined above in which there are microscopically visible differences between the chromosomes of different tissues are to be compared with the vastly larger number of cases where no differences between the chromosomes of different tissues have been detected by microscopy. Perhaps their chief significance is to indicate that there can be differences in the genetic make-up of different tissues of an individual and that the doctrine of the constancy of the genome during differentiation cannot be accepted as universally true.

Increases in the DNA content of the genome

Described above are a number of cases in which the genome of cells is apparently reduced during differentiation There are also a number of different situations in which the DNA content of nuclei may be increased during differentiation, so that the differentiated cells may have nuclei containing more DNA than the zygote.

Polyploidy

In the data of Tjio and Puck [136] discussed on p. 41 it can be seen that most cells in most tissues of the body contain the same number of chromosomes. In just a few instances, however, a multiple of this usual diploid number is found, as there is an increase in the number of complete sets of chromosomes, usually a doubling (tetraploidy). In the tissues examined by Tjio and Puck about 3 per cent of cells were polyploid, but the frequency of polyploidy is somewhat greater in liver than in other tissues. In the roots of plants polyploidy is common in the cells behind the growing point.

Polyteny

Another process which increases the DNA content of nuclei is polyteny, in which the chromatids replicate but do not separate, so that the number of strands of DNA in the chromosome is increased, perhaps a thousandfold. This greatly increases the width of the chromosomes and prevents the usual degree of coiling, so that the chromosomes also appear longer. As was mentioned on p. 41 such giant polytene chromosomes are to be found in various tissues of the larval stages of dipteran insects of many species. In such polytene chromosomes the heterochromatic regions usually increase to a smaller extent than the euchromatic regions.

In a few cases it has been found that the DNA content of some regions of the polytene chromosomes increases disproportionately compared with the remainder of the chromosomes and that this additional increase occurs in particular regions characteristic of the tissue and the stage of development. There is a superficial resemblance between such changes in the DNA content in particular regions of the chromosomes and the 'puffs' which are also to be seen in polytene chromosomes, and which display a particularly active synthesis of RNA around a certain region of the chromosome. These RNA puffs are further discussed in Chapter 5. The additional DNA production in certain regions is known by analogy as DNA puffing, and the phenomenon was described in the fly *Rhynchosciara* by Breuer and Pavan [16]. In a further study Rudkin and Corlette [112] showed that the DNA in the puff region at least doubled in amount, but they point out that since only a limited region of the chromosome is involved the total DNA content of the nucleus increases by less than 1 per cent.

Similar DNA puffs have been found in a few other flies, but at present little is known about the function of the increase in DNA content of the chromosome which occurs in the DNA puff. Much more is understood about the events involved in the DNA amplification which occurs in the oocyte.

Gene amplification in the oocyte

The **oocyte** is the stage during the formation of the female gametes in animals before the elimination of polar bodies and in which the DNA content is $4 \times C$. In many respects the oocyte is a specialized cell involved in the accumulation of the reserves of substances that will be utilized immediately after fertilization, and the gene amplification that is our present concern is involved in the production of the ribosomes that will be used in protein synthesis during the early stages of embryonic development.

The site of synthesis in the cell of the ribosomal RNA is the nucleolus, a microscopically distinguishable feature of the nucleus which is usually associated with a particular chromosomal location. By hybridization experiments in which radioactively labelled rRNA was annealed to DNA it has been established that the nucleolus region of the chromosome contains that DNA which is complementary to rRNA in base sequence and which is involved in rRNA synthesis [110, 148].

During the maturation of the oocyte in many species, including molluscs, insects, fish and amphibia, the number of nucleoli may be considerably increased, frequently accompanied by an appreciable increase in the amount of DNA in the nucleus, the additional DNA frequently being detached from the chromosomes. Many of these instances are reviewed by Gall [40]. A case that we may take as an example is the toad *Xenopus laevis*. The DNA responsible for the formation of the nucleolus (the **nucleolar organizer**) is normally represented once in each haploid set of chromosomes in this species, and so the tetraploid oocyte nucleus might be expected to contain four nucleoli. However, it appears that the nucleolar organizer region of the DNA is amplified by a factor of about 2,600 in the oocyte, so that about 1,500 nucleoli and 5,200 nucleolar cores are to be found in the oocyte nuclei [103]. Such an amplification of a specific DNA region can be seen as an adaptation which permits the oocyte to produce rRNA and ribosomes at a rate sufficient to ensure that the mature oocyte is plentifully supplied with the machinery for protein synthesis. For a review of work on the ribosomal RNA cistrons, the reader should turn to that by Birnstiel, Chipchase and Speirs [10].

Genetic systems outside the nucleus

For some time genetic experiments have indicated that some characters may be inherited in a way not explicable in terms of genes associated with the chromosomes in the nucleus, and it was suggested that the cytoplasm played a role in the determination of some heritable features. A number of reviews of cytoplasmic inheritance may be consulted, for example that by Wilkie [157]. More recently it has been established that some cytoplasmic organelles contain their own DNA which is utilized in the specification of RNA and probably protein molecules. The genetic systems of mitochondria and chloroplasts have been studied in detail, but cytoplasmic DNA is also found in centrioles and perhaps in other organelles. The significance of cytoplasmic factors in early embryonic development seems to be quite considerable and so it is perhaps appropriate to examine some of these extra-nuclear genetic systems in some detail.

Mitochondria

The presence of DNA in mitochondria has been established both by chemical studies on isolated mitochondria and by electron microscopy coupled with the use of specific enzymes to identify structures Work on the structure and function of this DNA is summarized in reviews by Granick and Gibor [45] and by Borst and Kroon [12]. The mitochondrial DNA is double-stranded and has a buoyant density which is frequently different from that of the nuclear DNA of the same species. In many cases it has been established by electron microscopy that the molecules of mitochondrial DNA are circular in form, as has been shown for example with mouse mitochondrial DNA. In this respect the mitochondrial DNA resembles bacterial DNA. However, the molecules of mitochondrial DNA are relatively small, and may consist of no more than 15,000 base pairs, which would not be sufficient to encode the information needed for the synthesis of the full range of mitochondrial proteins.

The mitochondria appear to contain all the machinery which is necessary for protein synthesis, for in addition to ribosomes, demonstrated by Rifkin, Wood and Luck [109] in *Neurospora* mitochondria, there is evidence that activating enzymes, transfer RNA and polymerizing enzymes are all present [142]. It has been established for some time that isolated mitochondria are capable of protein synthesis, but there has been some uncertainty about what proteins were synthesized. It appears that the mitochondria themselves are capable of forming their structural proteins [139, 6], while some at least of the mitochondrial enzymes and respiratory carriers are synthesized outside the mitochondria. Thus Kadenbach [69] has shown that cytochrome c is synthesized on cytoplasmic ribosomes and then transferred to the mitochondria. At this level, then, there is an interaction between the products of the genes of the nuclear and extra-nuclear systems. There is also genetic evidence that a number of mitochondrial enzymes are specified by structural genes which are in the nucleus and which behave in a normal Mendelian manner in hybridization experiments [32]. If the mitochondrial DNA is not required to code for all the mitochondrial catalytic proteins, then the amount of DNA present in the mitochondrion may be sufficient to specify the structural proteins as well as the mitochondrial RNA molecules which are apparently complementary with mitochondrial DNA. These RNA molecules include at least some of the tRNA molecules which differ in base sequence from the corresponding tRNA's of the extra-mitochondrial cytoplasm [91].

Summary

The commonly accepted model of the behaviour of genes during the growth and differentiation of an organism is that the genetic material is contained within the nucleus and is accurately transmitted during mitotic division so that all the cells in the body are supplied with the same genes in equal amounts. This is a simplification that we cannot fully accept. There are organisms in which not all cells have the same chromosomes; there are some cells in some organisms in which portions of chromosomal DNA are eliminated, or, alternatively, are increased in amount by gene amplification; there is evidence that during cell differentiation the nuclei may become limited in their capability of expressing all their genes; and there are systems of protein synthesis which are outside the direct control of the nuclear genes. These points must all be remembered when considering the mechanism of differentiation, but it nevertheless seems that in the majority of cases most genes are present in most types of cell. The problem of the mechanism of differentiation is how the activities of some of these genes are expressed at certain stages of development while other genes are suppressed.

5

GENE ACTIVITY AND
ITS REGULATION

We have seen from the arguments outlined in Chapter 3 that during cell differentiation there is a change in the complement of enzymes and other proteins in the cell, and that while some of this change may result from regulatory changes occurring after gene transcription, it nevertheless appears that the RNA content of cells changes, showing that differentiated cells differ in the products of their genes. Moreover, we saw from Chapter 4 that, though there may be important exceptions, most cells of an organism have the same complement of genetic information. From these lines of evidence we must conclude that cell differentiation involves changes in the proportion of the total genetic complement of the cell which is actually utilized in coding for RNA synthesis. The problem before us in this chapter is the assessment of possible mechanisms of control which could select and regulate the expression of genes.

The bacterial model

In bacteria it is frequently possible to change the pattern of enzymes produced by an organism by altering the conditions under which it is grown, the organism responding to its environment in a way which is clearly adaptive. For example, the bacterium *Escherichia coli*, when grown in the presence of lactose, produces a series of enzymes which enable it to absorb and utilize this sugar, hydrolysing it to glucose and galactose, which are then metabolized. Among the enzymes required for this is the one catalysing the hydrolysis of lactose, namely β-galactosidase (E.C.3.2.1.23). Lactose is one of a number of β-galactosides which **induce** the synthesis of this enzyme. A second enzyme which is induced simultaneously with β-galactosidase is β-galactoside permease, which promotes the absorption of lactose into the cell, while a third enzyme, β-galactoside acetyltransferase (E.C.2.3.1.18) is formed simultaneously, though its function in the organism is uncertain. If the **inducer**, lactose, is absent from the growth medium, the bacterium rapidly ceases the synthesis of these enzymes.

Other systems are known in bacteria in which specific substances, usually the end-products of metabolic pathways, may *inhibit* the synthesis of groups of enzymes. Such substances are known as **repressors**, while both inducers and repressors may be regarded as **effectors** of these enzyme systems.

Extensive study of the inducible enzymes of bacteria was carried out at the Institut Pasteur in Paris and in 1961 two workers at the institute, Jacob and Monod, proposed a theory which has proved invaluable in explaining the regulation of enzyme synthesis in bacteria [67]. The salient points of their model are set out in Fig. 28. They were

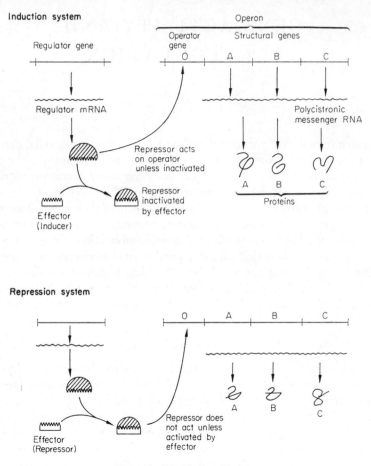

FIG. 28. The Operon Model

impressed by the way in which a series of enzymes could fluctuate in concentration in a parallel manner, and proposed that the **structural genes** which specify the template mRNA molecules for the group of enzymes are controlled as a unit. This group of structural genes was postulated to be under the control of an **operator gene**, the function of which is to prevent or permit the activity of the structural genes. A further gene is also proposed in the model: the **regulator gene**. This directs the synthesis of a **repressor substance** which interacts with the effector substance in the environment. The repressor is evidently a protein molecule which combines with the effector. The group of structural genes, together with the operator and regulator, is known as an **operon**. After combination with an inducer, the repressor no longer has the ability to bind to the operator gene, and so transcription of the structural genes is permitted. In the lactose system of *E. coli*, lactose combines with the repressor, inactivating it, so that the structural genes for β-galactosidase, β-galactoside acetyl-transferase, and β-galactoside permease can be transcribed.

An analogous system of control is proposed for those groups of enzymes the synthesis of which can be *inhibited* by a single inhibitor in the medium. In this model the regulator gene is envisaged as producing a protein molecule which in the presence of the effector *prevents* the structural genes from functioning.

The operon model has received considerable experimental support since it was propounded and provides a valuable explanation of the regulation of gene activity in bacteria, indicating how the range of enzymes synthesized may vary according to the circumstances of the organism, even though the genes present in the cell may remain the same. The system has been seen by many biologists as offering an analogy with differentiation in high organisms, and there has been a great temptation to suppose that the genes of eukaryotes might be regulated in a similar manner. However, there are reasons why we should be cautious about such an assumption.

The applicability of the operon theory to higher organisms

The great adaptability of bacteria to different environments and their capacity to change rapidly their metabolic activities in response to changing conditions are characteristics of these organisms which have presumably been largely responsible for their evolutionary success. Organisms such as the vertebrates, on the other hand, evolved a number of mechanisms which ensure that their cells are subject to a relatively constant environment, and in consequence there is probably less pressure to bring about rapid changes in the metabolic activity of the cells. There is thus no compelling reason to suppose that the mechanisms of metabolic regulation in higher organisms would be directly comparable with those of micro-organisms. If we turn to the characteristics of bacterial regulation explained by the operon model that are listed in Table 8 we find that these features are rarely seen in higher organisms.

TABLE 8. Characteristics of Regulation of Protein Synthesis According to the Operon Model

1.	Rapid and direct response to inducers and repressors such as enzyme substrates.
2.	Unstable messenger RNA (half-life of minutes)
3.	Groups of proteins regulated in a co-ordinated manner.
4.	Single messenger RNA molecule for all the proteins within the operon (polycistronic messengers).

A striking feature of bacteria is the way in which many enzymes are synthesized in response to the inducing action of their substrates. In higher organisms there are a number of clear cases where the synthesis of an enzyme can be specifically induced. Several of these cases are liver enzymes which are induced in response to hormones, such as tyrosine aminotransferase (E.C.2.6.1.5) induced by corticosteroids (see Chapter 6). However, there are few, if any, cases in higher organisms of enzymes being induced by their substrates. Moreover, the inductions which occur in response to hormones take place much more slowly than do bacterial enzyme inductions and the effect persists very much longer when the inducing influence is taken away.

Indeed, in differentiated cells the inducer of enzyme activity seems to act rather like a trigger, causing a response which continues after the removal of the inducer, while in bacterial cells the inducer must be continually present for the synthesis of the induced enzyme to persist. This difference is at least in part a reflexion of another important difference between the cells of higher organisms and bacteria, namely the much greater stability of the template for protein synthesis, the mRNA, in higher organisms. In bacteria mRNA decays with a half-life which is in the order of minutes, while in mammalian cells, for example, mRNA is often stable over periods of days, weeks, or even longer [107].

There are instances in higher organisms of groups of enzymes which are regulated together. Several of the enzymes of the urea cycle in the liver respond similarly to the hormone thyroxin and develop in parallel with each other (see p. 90). In the glycolytic pathway of carbohydrate metabolism certain enzymes, which are believed to exercise control over the pathway, are regulated together. Thus, glucose 6-phosphatase (E.C.3.1.3.9), hexosediphosphatase (E.C.3.1.3.11), pyruvate kinase (E.C.2.7.1.40) and pyruvate carboxylase (E.C.6.4.1.1) are induced by adrenocortical hormones and suppressed by insulin to similar extents, and have been said by Weber to constitute a **functional genic unit** [150].

One of the characteristics of the operon model is that the structural genes which are regulated together are genetically linked, lying alongside each other on the chromosome, and being transcribed together to produce a large mRNA molecule which codes for several proteins. Such messengers are said to be **polycistronic** as they represent several cistrons, or structural genes. There is very little evidence for the existence of such polycistronic messengers in the cells of higher organisms, or for the existence of close genetic linkage between groups of enzymes which are regulated together. This last point, however, must be seen against the background of the comparative poverty of accurate genetic mapping in higher organisms.

One of the features of the higher organisms is the presence of a discrete nucleus with a nuclear membrane, and it is this which places them in the group of **eukaryotes**. The **prokaryotes**, such as the bacteria, lack a nuclear membrane. Associated with this feature, we find that in the prokaryotes the RNA transcribed by the genes is frequently utilized directly in translation by the polysomes, while in eukaryotes the RNA is synthesized first as comparatively large molecules on the DNA of the chromosomes and may then be subject to considerable breakdown, apparently in rather specific processing steps, before it is released from the nucleus and used in protein synthesis [85]. This difference between the two great groups of organisms may also be of significance in their regulatory mechanisms.

It is clear that we cannot take the operon model of Jacob and Monod, which was specifically developed to explain bacterial gene regulation, and apply it without modification to the cells of higher organisms. Nor, on the other hand, should we reject it outright. There are some experimental data which are best interpreted by the supposition that specific regulator genes, as well as structural genes, are involved in enzyme control in higher organisms. One such instance concerns the enzyme tyrosine aminotransferase (E.C.2.6.1.5) (TAT) of mammalian liver, which has been investigated by Tomkins and his colleagues [137]. This enzyme is discussed in greater detail in Chapter 6, but the key observation which led to the proposal of a regulator gene in this system may be mentioned here. This stemmed from the use of the inhibitor **actinomycin D**. This substance inhibits the synthesis of mRNA by binding to DNA, and so usually has an inhibitory effect also on protein synthesis, the messenger which was present before actinomycin D treatment gradually decaying and not being replaced. The addition of actinomycin D to cultured liver cells might thus be expected to inhibit the synthesis of liver enzymes. However, under certain circumstances, TAT is increased in quantity following the administration of actinomycin D to liver, while under other circumstances treatment with actinomycin D permits the synthesis of TAT even in the absence of the adrenal steroid hormone which is normally required for the induction of the enzyme. Tomkins postulated that a regulatory gene must be present as well as the structural gene and that the mRNA for the regulator gene is less stable than that for the structural gene. In such a case, the presence of actinomycin D would be expected to inhibit the synthesis of the repressor first of all, and so the net effect could be a stimulatory one upon enzyme synthesis.

Somewhat similar mechanisms have been proposed to account for anomalous results with actinomycin D in connexion with tryptophan oxidase (E.C.1.13.1.12) [41] and with glutamine synthetase (E.C.6.3.1.2) in retina [1]. Nevertheless, although these experiments have been interpreted as evidence of the existence of controlling genes in addition to structural genes for some enzymes in higher organism, they require neither genetic linkage of regulator and structural gene, nor the regulation of several structural genes in a single operon.

Chromosome structure of eukaryotes

Table 9 contrasts the structure of the genetic apparatus of bacteria and of a typical mammalian cell. As mentioned earlier, the bacterial chromosome is not confined within a nuclear membrane, and there are several other important differences. The DNA content of cells of higher organisms is much greater than that of bacteria, and the chromosome is probably composed of DNA molecules coiled in rather a complex way so that a long portion of DNA is condensed into the chromosome. Moreover,

TABLE 9. Comparison of Genetic Apparatus in Bacteria and Higher Organisms

Bacteria	Higher organisms
No nuclear membrane	Nucleus enclosed in membrane except during nuclear division
Usually a single chromosome	Several to many chromosomes
DNA free of histones	DNA present as chromatin associated with histones and other proteins
About 10^{-16}g. DNA/cell	Up to about 10^{-12}g. DNA/cell

while *E. coli* has but one chromosome, eukaryotes have several and may have many. A further difference is that while the bacterial chromosome consists solely of DNA, eukaryote chromosomes also contain a number of different types of protein, both the basic proteins known as **histones** and the so-called **acidic** or **non-histone proteins**. In addition, there may be a species of RNA associated with the chromosomes. All these points of difference between the genes of bacteria and of higher organisms have been considered to be significant in gene regulation and so must make

61

us cautious in adopting a bacterial model of gene regulation when we come to consider plants and animals.

Chromatin

The complex of DNA, various proteins and RNA which makes up the chromosomes is known as **chromatin**, and by suitable methods it can be extracted from the nucleus and yet retain some of its functions. The most abundant proteins in the chromatin are the histones. They are a class of basic proteins, containing a high proportion of lysine and arginine, and they are roughly equivalent in amount to the DNA in the chromatin. A detailed study of the composition and function of isolated chromatin has been made by Bonner and his colleagues [11] and they quote ratios of histone to DNA ranging from 0·76 to 1·30 according to the species and tissue. The other proteins of the chromatin, which are not basic in character but rather acidic, are much more variable in their proportions in the chromatin, with ratios relative to DNA ranging from 0·10 to 1·04. The amount of RNA that is associated with the chromatin is smaller in amount. The weight of this RNA relative to DNA may be as low as 0·007, while the highest proportion quoted by Bonner is 0·26. These variations in the composition of chromatin from different sources may be of great significance in the regulation of the function of the chromosomes as we shall see later, but first we must look at the major companion to the DNA in the chromosomes, the histones.

The histones

As long ago as 1950 Stedman and Stedman [126] drew attention to the varying proportions of different types of histone in different tissues, and speculated that they might in some way be involved with tissue differentiation. Since then much work has been done to characterize these proteins more fully and to develop methods of fractionating them. Nevertheless, the number of different types of histone found remains small, only five to eight different types being detectable in most individual plants or animals, while there are very marked similarities in the sequence of amino acids that make up the polypeptide chains in histones derived from very different species. These similarities in protein structure suggest that the histones play a very significant role in a wide range of plants and animals and that any changes in structure brought about by random mutations are so disadvantageous to the individual affected as to be eliminated by natural selection. On the other hand the limited range of variation of molecular structure that has been found so far among the histones makes it difficult to envisage them having a function which is specific for a given tissue. If they were to be tissue-specific, then we would expect all the different types of cell in the body to have a different complement of histones. Even so, many studies of the role of histones in isolated chromatin preparations point to an important regulatory role for these proteins.

Bonner and his co-workers found that when isolated chromatin was incubated in the presence of nucleotide precursors of RNA and with the enzyme RNA polymerase (RNA nucleotidyltransferase, E.C.2.7.7.6) which had been prepared from bacteria, then the chromatin would direct the synthesis of RNA. In this reaction the DNA of the chromosomes acts as a template and the bases in the RNA are aligned in a specific sequence by pairing with the complementary bases in the DNA. The rate at which the chromatin could act as a template for RNA synthesis was not however as great as that of pure DNA. When the histones were removed from the chromatin by extraction with dilute hydrochloric acid, there was a marked increase in the ability of the preparation to act as a template for RNA synthesis. It appeared, then, that the histones restricted the DNA in its activity of directing RNA synthesis. The type of RNA that is synthesized by a chromatin template was investigated using the RNA/DNA hybridization technique by Paul and Gilmour [100]. They found that the range of types of RNA synthesized was greater with DNA from which the histones had been removed than with the chromatin itself. Bonner's group of workers also carried out experiments in which histones were added back to purified DNA, so reconstituting the chromatin to some extent. They interpreted their findings as an indication that the type of RNA formed depended upon the type of histones that were present.

Important experiments have been carried out using nucleic acid hybridization techniques to show that the types of RNA which are synthesized when isolated chromatin acts as a template are comparable to those which are normally synthesized by the tissues from which the chromatin was isolated. This type of work began with that of Paul and Gilmour and a more detailed investigation was that of Smith, Church and McCarthy [122]. These experiments used RNA isolated from various tissues as a competitor for hybridization of DNA with radioactively labelled RNA which had been synthesized on templates of isolated chromatin. As is seen in Fig. 29 the RNA

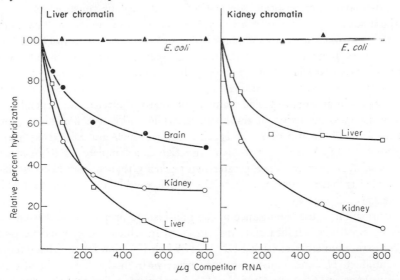

FIG. 29. Tissue Specificity of Template Activity of Isolated Chromatin Competition by tissue RNA with hybridization of radioactive RNA synthesized with an isolated mouse liver chromatin template and mouse RNA polymerase. Hybridization was with mouse DNA.

From Smith, K.D., Church, R.B. and McCarthy, B.J.: Template specificity of isolated chromatin. *Biochemistry* 8, 4271–77 (1969). Copyright (1969) by the American Chemical Society. Reprinted by permission of the copyright owner.

from liver was the most effective competitor when challenging RNA which was synthesized on a liver chromatin template, other tissue RNA preparations being less effective competitors, and RNA from bacteria having practically no competitive effect. Similarly, kidney RNA competed most effectively with RNA synthesized on a kidney chromatin template and brain RNA competed with the product derived from brain chromatin. Besides demonstrating the specificity of action of the isolated chromatin, these experiments provide further confirmation of the differences in the RNA compositions of the different organs of the body.

It appears that histones repress the activity of DNA in acting as a template for RNA synthesis in isolated chromatin, and that the chromatin of different tissues acts to synthesize a different range of RNA molecules. The experiments described above are consistent with the view that tissue specificity is achieved by restricting the amount of the genome that is able to be transcribed into RNA.

Changes in the template activity of chromatin from different stages of embryonic development have been demonstrated by other workers such as Flickinger and his colleagues [39], while a change in the template activity of liver chromatin during the metamorphosis of the frog is discussed on p. 90.

Although the histones constitute an important part of the chromatin of the cell nucleus, we should not overlook the function of the other constituents, particularly in view of the fact that the limited specificity apparently possible with a small range of types of histone molecule poses theoretical problems about how the histones could recognize which of the genes they should repress. The significance of the non-histone proteins has been emphasized by some experiments of Paul and Gilmour [101]. They attempted to reconstitute chromatin from its various constituents and found that there was a normal template activity, that is one similar in its products to that of the intact cell, when histones *and* non-histone protein were added to DNA, but that when histones alone were added there was a total inhibition of template activity. Paul [98] has summarized the conclusions from these experiments by stating that differentiation involves the masking of certain specific sequences of DNA in the chromosomes in an organ-specific manner and that the mechanism depends upon a non-specific masking action of the histones, while within the non-histone protein fraction of the chromatin there are specific factors which unmask certain genes, so permitting the DNA in these regions to function as a template for RNA synthesis. The non-histone proteins show greater heterogeneity than do the histones, so that they appear to have a potential specificity which would be expected of regulatory proteins [78].

Confirmation of the significance of the non-histone proteins comes from experiments by Spelsberg, Hnilica and Ansevin [123] in which the synthetic activity of templates was studied, again using DNA-RNA hybridization for the characterization of the product. Histones and non-histone proteins from the chromatin of different tissues, thymus and liver, were added to DNA and it was found that the type of RNA synthesized did not depend on the origin of the histone, but only on that of the non-histone protein. Kamiyama and Wang have recently taken this type of experiment a step further by showing that the RNA produced in response to the addition of non-histone protein is capable of acting as a template for polypeptide synthesis and that following addition of non-histone protein to the system different polypeptides are formed compared with those synthesized in the controls [70].

These views about the significance of the non-histone protein fraction are not

64

shared by all workers in this field, and controversy may be expected to rage for some time before there is any generally accepted view of the mechanism by which the template activity of DNA is controlled in the cell. Perhaps the chief value of the account of recent work in connexion with this problem is as an illustration of the approaches and ideas that were in fashion at the time of writing. These molecular models of chromosome function are somewhat remote from direct observation. We should now see what more direct evidence we have of the way in which the chromosomes may function during development.

Polytene chromosomes and puffing

In the polytene chromosomes of some flies (Diptera) we have an example of the regulation of gene expression which can be visualized more directly than any other system and from which we can gain useful ideas about mechanisms of gene regulation. The general structure of polytene chromosomes was mentioned in Chapter 4 and we saw that by the multiplication of the number of strands of the chromosomes accompanied by the maintenance of the alignment of corresponding regions of the strands it becomes possible to see by the use of the light microscope structures which can be related to the genes that are present on the chromosomes. In Chapter 4 emphasis was placed on the similarity of position and sequence of bands in different tissues. However, the dimensions of the bands and the intensity with which they stain with certain histological dyes does vary, not only between tissues but also at different times within the same tissue. At various times certain bands in the chromosome may become enlarged into what are known as **puffs** (Fig. 30). Certain very large puffs have at times been given a different name—**Balbiani rings**, but these are merely a more extreme manifestation of the same phenomenon. The puffs are transient structures within the chromosome, and during the development of a tissue there is a sequence of formation and disappearance of puffs at different genetic sites on the chromosome. This sequence is characteristic of the tissue and also depends upon the metabolic state of the insect (Fig. 31). In particular, the processes of moulting when the old cuticle is shed and a new one comes to the surface, and also the more extreme change of pupation, are accompanied by the puffing of particular sites on certain chromosomes [3].

Microscopic studies, together with autoradiographic estimation of the uptake of radioactive precursors of proteins and nucleic acids, have shown that the puffing of chromosomes is accompanied by the rapid synthesis of RNA, though in most cases the amount of DNA in the region of the puff remains constant. This synthesis of RNA is in agreement with the view that puffing represents the activity of the gene as it is transcribed, and attempts to characterize the RNA that is synthesized as a gene undergoes puffing have indicated that it has some of the properties of mRNA [9].

The moulting of insects is regulated by a system of hormones, one of which, **ecdysone** (Fig. 32), brings about the changes in the insect cuticle leading to the sloughing off of the old cuticle and the development of the new exoskeleton. Characteristic enzymes are formed as this moulting process takes place, and these enzymatic changes can be brought about by treatment of the insect with ecdysone. This hormone also gives rise to the patterns of chromosome puffing which are characteristic of the normal moulting process. Such changes in puffing pattern can be brought about by addition of ecdysone to certain organs of the insect maintained in tissue culture, and changes can even be seen when individual cells are treated with the hormone [26].

The puffing of insect polytene chromosomes provides us with a system in which

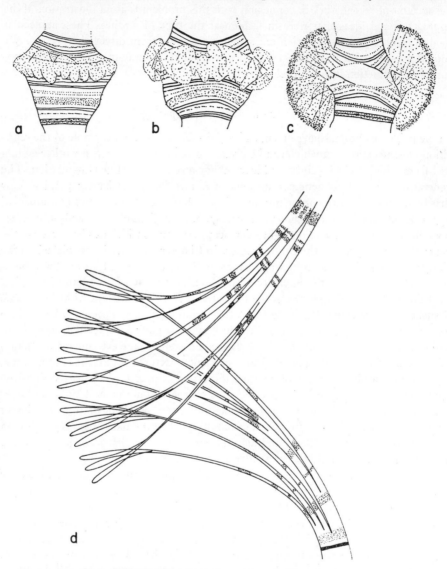

FIG. 30. Structure of a Puffed Site of a Polytene Chromosome
(a to c) Different degrees of puffing at the same site
(d) Diagrammatic representation of part of a helically wound bundle of chromonemata at the location of a fully developed puff.

(a to c) from Beermann, W.: Control of differentiation at the chromosomal level. *Journal of Experimental Zoology* **157**, 49–61 (1964). Copyright (1964) by The Wistar Press. (d) From Beermann, W.: Nuclear differentiation and functional morphology of chromosomes. *Cold Spring Harbor Symposia on Quantitative Biology* **21**, 217–30 (1956). Copyright (1956) by Cold Spring Harbor Laboratory.

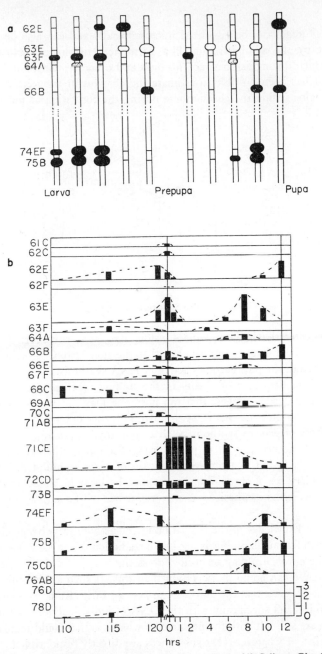

FIG. 31. The Sequence of Puffing at Different Loci on a *Drosophila* Salivary Gland Chromosome.

(a) The sequence of appearance and regression of seven puffs on chromosome arm 3L.

(b) Histograms showing the changes in mean puff size of puffs on the same chromosome during the late third instar larva and prepupal stages.

From Ashburner, M.: Patterns of puffing activity in the salivary gland chromosomes of Drosophila. I. Autosomal puffing patterns in the laboratory stock of Drosophila melanogaster. *Chromosoma* (Berl.) **21**, 398–28 (1967). Berlin-Heidelberg-New York: Springer.

the activity of genes can be visualized microscopically and can also be regulated to some extent by the administration of a hormone. Histological stains have been developed which distinguish between histones and other proteins, and by the use of such stains it has been possible to estimate the changes in the relative amounts of different proteins within the chromosome during the formation of puffs. It appears

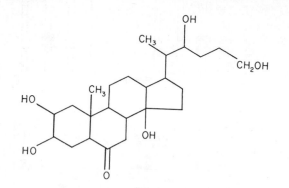

FIG. 32. The Structure of Ecdysone

that there is no alteration in the amount of histone associated with a certain region of the chromosome as it undergoes puffing, but that the non-histone protein increases in amount just before the puff becomes active in RNA synthesis.

Studies of the sequence of events in the puffing of chromosomes induced by treatment with ecdysone suggest that the hormone may initiate the puffing of one set of chromosome sites and that these, in turn, influence other sites to begin puffing. When inhibitors of protein synthesis, such as puromycin, are applied to the system it is found that the primary set of puffs, brought about by the action of ecdysone, takes place normally but that the secondary set of puffs is inhibited. This suggests that the secondary set of chromosome puffs may be brought about by the action of some protein which is synthesized as a consequence of the activity of the primary puffs.

Our present knowledge of the regulation of the activity of polytene chromosomes of insects is still fragmentary, but is not in contradiction of the view that non-histone proteins of the chromatin may act to derepress the template activity of the DNA of certain genes. We can be no more certain than this. If such a mechanism does take place we must presume that the non-histone protein varies from one gene to another and is able to interact specifically with the DNA of individual genes or small groups of genes. We are still left with the problem of what it is that could regulate the non-histone proteins, or whatever other substances are directly responsible for controlling DNA template activity. Thus the problem of the more remote causes of the differential stimulation, or derepression, of gene activity in differentiating cells remains with us.

The influence of the cytoplasm on genetic activity

A large body of knowledge derived from classical experimental embryology shows that in many early developing embryos certain regions of the egg give rise to particular organs and tissues. This knowledge is epitomized by the construction of embryonic 'fate maps' in which the normal end-products of particular regions of the egg are recorded. This data tell us that the cytoplasmic structure of the fertilized egg has encoded within it, in some form, the information that directs the gene activity of nuclei at later stages of development. A more specific instance of this is seen in experiments on the development of the nematode worm *Ascaris* in which certain cells form the gonads, or germ-line, retaining their full chromosomal complement, while other cells, forming the somatic line, suffer the loss of portions of the chromosome (see Chapter 4). It has been demonstrated that the fate of the nuclei of the developing *Ascaris* depends upon which region of the egg cytoplasm they occupy. Experimental movement of the nucleus to a different region of the cytoplasm can alter the development of the chromosomes [147]. Classical experiments on embryonic development provide us with the clue that the cytoplasm can regulate the activity of the nucleus to some extent. More recent biochemical studies have elaborated this idea.

The most direct manifestation of the activity of genes is the synthesis of RNA. As we have seen in earlier sections of this book, the mRNA is of especial interest in the regulation of cell differentiation, but other species of RNA are of great importance and are sometimes much easier to study experimentally. Ribosomal RNA (rRNA) and transfer RNA (tRNA) are classes of RNA which are direct gene products produced in larger amounts than any mRNA and so they provide valuable markers of gene activity during development. The normal pattern of synthesis of different classes of RNA has been established in a number of species of animals. For example, the times of synthesis of rRNA and tRNA in the early embryos of the toad *Xenopus laevis* have been established by D D Brown and his colleagues (see review by Gurdon [52]). During the growth of the oocyte there is some synthesis of all classes of RNA, but this synthesis slows down as the oocyte becomes mature and there is virtually no RNA synthetic activity in the unfertilized egg. Ribosomal RNA synthesis remains in abeyance until gastrulation, tRNA synthesis shows a burst of activity at the blastula stage, while the synthesis of the class of RNA which is heterogeneous in size and which probably represents mRNA shows relatively high levels quite soon after fertilization and in the blastula stage. This is shown in Fig. 33.

Nuclear transplantation and RNA synthesis

The technique of nuclear transplantation, mentioned in Chapter 4, has been applied by Gurdon and his colleagues to an investigation of the factors regulating RNA synthesis [54]. The results of these experiments indicate that the cytoplasm has a profound effect upon the RNA synthesis of the nucleus. Gurdon found, for example, that when a nucleus was removed from a late neurula of *Xenopus laevis*, in which it would be synthesizing rRNA, and transplanted into the cytoplasm of an uncleaved egg, then there was an inhibition of rRNA synthesis. Similarly, a nucleus from the gastrula stage, normally synthesizing tRNA, was found to stop such synthesis on implantation into egg cytoplasm. Now it might seem that these inhibitions could be due to damage to the nucleus during transplantation, but the progeny produced by mitosis of the transplanted nuclei did show their normal synthetic activity when the embryo had developed to the appropriate stage. Clearly, no irreversible change had occurred in the transplanted nuclei. Moreover, the changes brought about by trans-plantation need not necessarily be inhibitions. For example, mid-blastula nuclei which

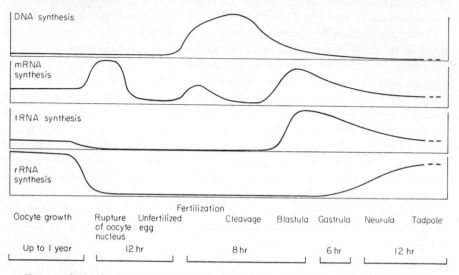

FIG. 33. Approximate Rates of Synthesis of the Main Classes of Nucleic Acid During Normal Amphibian Development

From Gurdon, J.B.: Nucleic acid synthesis in embryos and its bearing on cell differen-tiation. *Essays in Biochemistry* **4**, 25–68 (1968). Copyright (1968) by the Biochemical Society.

did not synthesize RNA, could be implanted in the cytoplasm of the oocyte, and in these circumstances RNA synthesis was stimulated.

These experiments making use of nuclear transplantation show that the pattern of RNA synthesis in the nucleus is influenced by the developmental state of the cyto-plasm. Unfortunately the technique of nuclear transplantation is limited in its appli-cation because of the necessity to use large cells as the cytoplasmic host for the trans-planted nuclei, so that only the influence of oocyte or egg cytoplasm can be studied. Another approach yielding similar kinds of information can be made using the technique of cell fusion.

Cell fusion

When cells are infected with certain viruses they may clump together and this may lead eventually to cell fusion, so that more than one nucleus may be embedded in a single continuous mass of cytoplasm. Such multinucleate cells are found in many virus-infected tissues. In 1965 Harris and Watkins showed that killed virus particles would cause fusions between cells of different species and that the multinucleate cells so formed could live. By the use of autoradiography following the addition of either radioactive uridine or radioactive thymidine it was possible to determine whether RNA and DNA were synthesized in the nuclei [58].

The lymphocytes that circulate in the blood only very rarely show detectable RNA synthesis, while the nuclei of the human tumour cell line, HeLa cells, show a pro-nounced and continuous RNA synthesis. After HeLa cells and rat lymphocytes have been fused with the aid of virus it is still possible to distinguish microscopically between the types of nuclei, according to the cell from which they are derived. By autoradiography it was found that both types of nuclei, lymphocyte and HeLa cell, were active in RNA synthesis after cell fusion. The lymphocyte nucleus is evidently stimulated to RNA synthesis by its new cytoplasmic environment. A similar result was obtained with experiments using the erythrocyte of the hen. Although the erythrocyte of birds retains its nucleus, unlike that of mammals, this nucleus shows little activity and synthesizes neither RNA nor DNA. However, after fusion with HeLa cells the nucleus derived from a hen erythrocyte showed significant synthesis of both RNA and DNA. It is necessary to establish that the rather unusual circumstances of cell fusion are not responsible for some unspecific effect which stimulates all nucleic acid synthesis indiscriminately. This idea can be discounted by an experiment in which rabbit macrophages and rat lymphocytes, neither of which synthesize DNA, were fused together. No stimulation of DNA synthesis was found in this instance [58].

The signal from cytoplasm to nucleus

Both the nuclear transplantation and the cell fusion experiments show that the synthesis of RNA and DNA is influenced by the cytoplasm. Moreover, the cell fusion results indicate that this cytoplasmic effect is not due to an inhibitory signal, but rather that there is a positive signal from cytoplasm to nucleus: nuclei which would otherwise be inactive can be stimulated by the presence of some cytoplasm which supports nucleic acid synthesis, while active nuclei are in no way inhibited by that cytoplasm which does not normally support RNA or DNA synthesis. We conclude then that the cytoplasm sends to the nucleus a signal which stimulates genetic activity. Gurdon and Woodland have indeed demonstrated that before nuclei increase their synthetic activity after being implanted into stimulatory cytoplasm they undergo a marked increase in volume, which they suppose is due to proteins entering the nucleus. Some experiments with labelled proteins also indicate that such movement may take place [2], but as yet we cannot say that the signal that passes from cytoplasm to nucleus and that stimulates gene activity is certainly a protein.

Embryonic induction

It is not proposed to discuss embryonic induction extensively since many textbooks are available for this topic [115, 56]. The classical cases of embryonic induction were described by Spemann and Mangold in 1924. They demonstrated that the normal development of the nervous system of the early embryo depends upon contact between the roof of the archenteron and the ectoderm of the embryo. Without such contact the primary axis of the embryo containing notochord and neural tube is not formed. On the other hand, the grafting of an additional dorsal blastopore lip into an embryo brought about the production of an additional embryonic axis. In this way some ectodermal tissue was made to alter its normal path of development and give rise to neural tissue.

Since Spemann's time many systems of induction have been studied and in some cases very clear-cut changes in the developmental fate of tissues have been brought about by altering experimentally the nature of the tissue from which the inductive stimulus emanates. For example, during the normal development of the eye the ectodermal epithelium of the embryo gives rise to the cornea, in response to an induction by the lens. However, if the lens is removed from the embryonic chick eye and mesenchymal tissue from the foot region is implanted instead, then scales can be induced to form instead of cornea; while mesenchyme from certain prospective dermis regions can bring about the induction of feather formation in place of the cornea [29].

Something of the nature of the inductive process has been revealed by experiments carried out by Grobstein on the differentiation of the pancreas. He showed that epithelium which in the whole animal would give rise to pancreas did not differentiate if grown alone in tissue culture, but that when mesenchyme tissue was allowed to come into contact with it the epithelium would differentiate and show features of pancreas morphology and biochemistry. By setting up the tissue culture system with a porous membrane between the epithelium and the mesenchyme Grobstein showed that the influence of the mesenchyme was able to act through the membrane. Although cells were not able to penetrate through the membrane, the presence of pores of about 0.8μ in diameter permitted the transfer of quite large molecules. If a membrane made of cellophane was used, which was permeable only to small molecules, then induction did not take place. These experiments suggest that the transfer of macromolecules between tissues is essential to induction [48].

In some systems of tissue interaction the induced tissue can develop along one of a number of different pathways according to the nature of the inducing tissue. For instance, undifferentiated ectoderm can be made to develop into either neural tube or lens according to the underlying tissue. This strongly suggests that in some cases, at least, the signal passing during induction must convey information, rather than acting as a mere trigger eliciting a response which is already inherent in the induced tissue.

Many attempts have been made to determine the nature of the inducing signal, and the primary embryonic inducer which stimulates the formation of the neural tube from the embryonic ectoderm has been studied for many years. The work of Tiedemann and his colleagues has indicated that this inducer is probably a protein, while another inducer which leads to the differentiation of mesodermal structures such as notochord, muscle and renal tubules has been further purified and shown to be a protein with a molecular weight of 25,000 to 30,000 [134]. Preliminary experiments indicate that the function of the inducer involves its uptake by the induced cells. As yet there is no evidence about whether the inducer acts on the cytoplasm or directly on the nucleus of the target cells.

Hormones and differentiation

Just as there is evidence that at least some embryonic inducers may convey information which determines the course of differentiation, so we find that certain hormones clearly direct development along a particular pathway which would otherwise not be taken. This can be well illustrated by the influence of the sex hormones on the development of gonads and accessory sex organs. Although sex is determined in

72

higher organisms by a genetic mechanism involving specific sex chromosomes, the embryo of vertebrates passes through a phase in which the gonads, ducts and genitalia are identical in male and female embryos. In the course of normal development certain structures become enlarged and others diminish according to the sex of the embryo. This is illustrated by Fig. 34. If an embryo is treated with hormones appropriate to

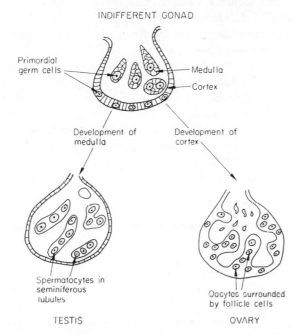

FIG. 34. Differentiation of Gonads from the Indifferent Immature State
After Balinsky, B.I. *An Introduction to Embryology*. Philadelphia: Saunders.

the opposite sex it is sometimes possible to alter the course of development so that the anatomy of the embryo reveals a sex opposite to that indicated by the genotype. For example, an opossum with a chromosomal complement appropriate to a male can be treated with sex hormones so as to develop into a creature that is phenotypically female [20]. Such experiments do not, of course, represent the normal course of development, but they indicate that the direction of development can be crucially influenced by information extrinsic to the developing tissues.

A well-known example of the effect of hormones in development is in amphibian metamorphosis. The many changes, both morphological and biochemical, which take place when a tadpole becomes a frog can be brought about prematurely by administration of thyroxin. These changes include resorption of the tail, stimulation of the growth of the hind limbs, bulging of the eyes, shortening of the gut and induction in the liver of the enzymes catalysing the synthesis of urea (see p. 90). Thus a single substance brings about a wide range of changes, and the same type of tissue in different locations changes in different ways. For example, tail muscle undergoes degenerative changes under the influence of thyroxin while limb muscle does not. Moreover, tail muscle grafted on to the flank of the animal will show these degenerative changes after hormone treatment even though the flank muscle all around will not. The behaviour of a tissue in response to a hormone does not depend just on the hormone but also on the state of the tissue which must be **competent** to respond. The same phenomenon of competence can be seen also with respect to embryonic inducers. There is frequently only a limited period in development when a tissue may respond to an inducing influence. Further examples of competent periods in hormone action are mentioned in the account of liver differentiation in Chapter 6.

Detailed biochemical studies of the mechanism of action of hormones have confirmed the idea that some hormones, at least, exert their action by influencing the activity of genes. We have already seen (p. 65) how ecdysone acts at the gene level in insect moulting. Other examples can be found in higher organisms. It has been demonstrated that the hormone testosterone stimulates RNA synthesis in the prostate gland, leading to the accumulation of RNA with an altered base composition. These changes can be inhibited by actinomycin D [76]. Oestrogens have also been shown to affect the transcription of genes. Hamilton [57] has reviewed experiments which show that oestradiol binds to the chromatin of the nuclei of rat uterus tissue and that this is followed by the stimulation of RNA synthesis. More recently, experiments have indicated that oestrogen acts by inducing in the nucleus the synthesis of a specific non-histone protein which has a regulatory function [132]. A more direct demonstration of the influence of hormones at the gene level is seen in experiments described by O'Malley and colleagues. They were able to isolate the mRNA which directs the synthesis of the protein avidin in the oviduct of the chick and they showed that treatment of young chicks with oestrogen caused the oviduct to produce this messenger [96].

Though cells and tissues receive information from outside in the form of hormones and inducers, yet their response depends on something intrinsic to the cells, and, we may guess, intrinsic to the state of the cell nucleus. The regulation of the genetic activity of cells seems to depend upon a complex interaction of cell nucleus, cytoplasm and cellular environment.

Summary

Theories concerning the regulation of genetic activity developed in connexion with bacteria may be of interest to studies of differentiation but they cannot be applied without modification. Experimental evidence does, however, support the view that in differentiated cells some genes may control the activity of others.

The ability of DNA to act as a template for RNA synthesis is influenced by the type of protein bound to the DNA in the chromatin. Non-histone proteins (acidic proteins) may play a particularly important role in specific regulation of template activity.

In the RNA puffs of dipteran polytene chromosomes differential gene activity may be visualized and the influence of hormones on this activity demonstrated.

The activity of the DNA in nuclei is influenced by the cytoplasm, probably as a result of a positive signal passing into the nucleus, while other external influences on gene activity, such as some inducers and hormones, may also act via the mediation of the cytoplasm.

6

SOME SYSTEMS OF
CYTODIFFERENTIATION

In previous chapters some of the general problems of cell differentiation have been set out. In this chapter a number of individual systems of differentiating cells will be examined by way of illustration and in order to seek some generalizations about the process of cytodifferentiation.

THE ERYTHROCYTE

Erythrocytes are produced at a high rate throughout the life of vertebrates and the site of their production may change during the course of development. In mammals the earliest location of the formation of red blood cells, or **erythropoiesis** as it is known, is the yolk sac of the embryo, then the liver, the spleen and finally the bone marrow take over in turn as the major sites of formation of red blood cells. In man this role of the bone marrow begins in the seventh month of gestation.

The outstanding biochemical feature of the erythrocyte is its possession of a very high concentration of haemoglobin and only very small amounts of other proteins, chiefly the structural proteins of the cell membrane. The morphological features of the mature mammalian erythrocyte are also very characteristic, the absence of a nucleus, mitochondria, ribosomes and Golgi apparatus being notable. Thus there is a degree of specialization of function in the erythrocyte which is matched by few other cells. The erythrocyte is the end of a line of differentiation which starts with a cell known as a **haemocytoblast**, and histologists have given names to several of the stages between this starting point and the mature erythrocyte. Some of these are indicated in Fig. 35, which also indicates some of the changes in structure that occur as differentiation proceeds. Cell division can occur in the early stages of differentiation, and, of the daughter cells produced by mitosis, one, the stem cell, remains undifferentiated and capable of further division, while the other daughter cell begins to differentiate. The relative size of the nucleus diminishes and the number of polysomes in the cell decreases as development continues. In late embryonic and adult mammals the nucleus is lost before the ribosomes cease to function, while in birds the nucleus, though not extruded from the cell, is nevertheless completely inactivated. The reticulocyte stage is that at which the nucleus is lost or non-functional while functioning ribosomes still remain. This shows that in the reticulocyte mRNA is still present in the cytoplasm after it has ceased to be produced by the nucleus.

F   75

General reviews on erythrocyte differentiation are given by Goldwasser [44], Marks and Kovach [82] and Wilt [159].

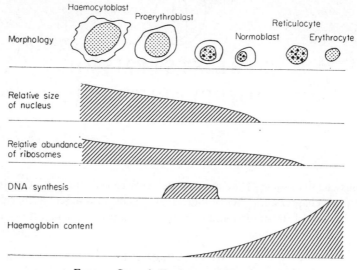

FIG. 35. Stages in Erythrocyte Differentiation

Haemoglobin structure

The normal haemoglobin molecule consists of four polypeptide chains of two different types, to each of which is attached a haem group, the site of the binding of oxygen. The most abundant type of haemoglobin in adult humans, known as haemoglobin A, contains polypeptide chains of the types known as α-globin and β-globin, and the composition of the tetramer may be represented by the formula $\alpha_2\beta_2$. The complete amino acid sequence of both of these polypeptide chains is known and the spatial structure of the complete molecule has been found by X-ray crystallographic studies. Details will be found in many textbooks and reviews [66]. In the human foetus another type of haemoglobin, known as haemoglobin F, is found in which there is a different polypeptide chain, known as γ-globin, with some marked similarities in structure to β-globin. Table 10 lists the composition of several different types of haemoglobin that are found in man. All have four polypeptide chains and four haem groups.

TABLE 10. The Subunit Composition of Human Haemoglobins

Haemoglobin A	The bulk of normal adult haemoglobin	$\alpha_2\beta_2$
Haemoglobin A$_2$	About 2·5 per cent of the haemoglobin of a normal adult	$\alpha_2\delta_2$
Embryonic haemoglobin	Present in first months of gestation	$\alpha_2\varepsilon_2$
Haemoglobin F	Predominating in the foetus and immediately after birth	$\alpha_2\gamma_2$

Regulation of haemoglobin synthesis

The persistence of protein synthesis in the reticulocyte after the loss of nuclear activity indicates that the mRNA molecules for globin chains are relatively stable.

76

Further evidence of this comes from the work of Wilt [159] who studied the formation of haemoglobin in cultured chick embryos. He showed that haemoglobin could be detected after the blastodisc (the developing cells of the egg which form the embryo) had been incubated for 30 hours and that treatment of the embryos with the inhibitor actinomycin D at this stage, though it inhibited RNA synthesis, did not inhibit the synthesis of haemoglobin. If the treatment with actinomycin D was carried out after 24 hours of incubation, causing a complete inhibition of RNA synthesis, haemoglobin still appeared 6 hours later, at 30 hours of incubation. However, if actinomycin D was used before 24 hours of incubation, then haemoglobin formation was prevented. The mRNA needed for haemoglobin synthesis appears to be synthesized at least 6 hours before the globin chains are formed. These experiments indicate that the presence in the cells of the mRNA coding for globin is not in itself sufficient to bring about the synthesis of haemoglobin. Some regulatory process seems to operate to prevent the synthesis of globin even after the globin genes have been transcribed to produce mRNA. This regulatory mechanism must act during the translational stages of protein synthesis, and indeed this is one of the clearest instances of the translational control of protein synthesis that we have in embryonic development.

Since haemoglobin contains four haem prosthetic groups per molecule, then the synthesis of functional haemoglobin involves not only the formation of the polypeptide chains but also the elaboration and attachment of these haem groups. There is clear evidence that haemoglobin synthesis is, in part, controlled via the metabolic pathway of haem synthesis. A key enzyme is that catalysing the synthesis of δ-aminolaevulinic acid from glycine and succinate, and known as δ-aminolaevulinate synthetase.

Physiologists are well aware that bleeding or low oxygen pressure stimulates the formation of new red blood cells. This occurs for instance in adaptation to high altitudes, and the stimulation is brought about by the action of a substance called **erythropoietin**. The apparent action of erythropoietin is to increase the activity of δ-aminolaevulinate synthetase, probably by increasing the rate of synthesis of this enzyme [44]. In this way haem synthesis is increased. In cell-free systems prepared from reticulocytes and which are able to synthesize haemoglobin, the addition of haem increases the rate of globin synthesis and in this way the levels of haem and globin within the cell are kept in balance. Although the mechanism of the stimulation by haem is not known with certainty, it has been proposed that the globin formed in the absence of haem is unstable and becomes modified in the cells so that it acts as an inhibitor of the initiation of the synthesis of globin chains by polysomes [108].

It appears that erythropoietin stimulates the synthesis of haemoglobin by first acting on the synthesis of δ-aminolaevulinate synthetase, which leads to increased haem formation and hence to the stabilization of nascent globin chains. Fig. 36 indicates this sequence of events. Certain steroids also act on haemoglobin synthesis by stimulating δ-aminolaevulinate synthetase and haem synthesis [74].

One further effect of haem should also be mentioned. This is its effect on the activity of δ-aminolaevulinate synthetase which it diminishes. There is thus a negative feedback loop which tends to maintain a constant level of haem since its accumulation leads to a reduction in its rate of synthesis.

Although the detailed molecular mechanisms of these interactions in haemoglobin synthesis are not yet worked out they nevertheless provide good examples of the types of control mechanisms which might be involved in the synthesis of many proteins in the cell.

There is not yet any sure knowledge of the way in which translational control is exerted during the synthesis of globin in the chick embryo, but it is possible that one,

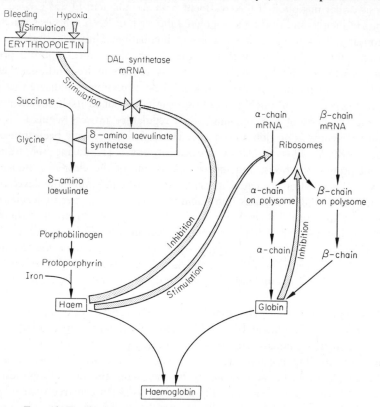

FIG. 36. Feedback Loops in the Regulation of Haemoglobin Synthesis

at least, of the tRNA molecules needed to translate globin message may be absent in the early embryo so that globin synthesis can only begin when this tRNA becomes available. Thus the activity of a gene for tRNA production would also be involved in haemoglobin synthesis.

A further problem to be considered in haemoglobin production is the mechanism whereby the proportions of α- and β-chains are matched so that the completed molecule contains equal numbers of the two types of chain. It has been suggested that the presence of a completed β-globin chain is necessary for α-chains to be released from the polysomes [159].

Changes in haemoglobin composition during development

In many animals it has been found that different types of haemoglobin are formed at different stages of development. In the frog, tadpole haemoglobin differs in composition from that of the adult, a change in the pattern of synthesis of haemoglobin being among the many changes involved in metamorphosis. As was mentioned above there are developmental changes in the polypeptide chains present in haemoglobin in man, and these are represented in Fig. 37. In the early embryo, up to 3 months of gestation,

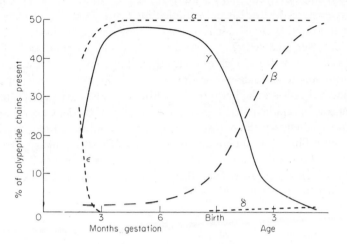

FIG. 37. Changes in the Subunit Composition of Haemoglobin
From Huehns, E.R., Dance, N., Beaven, G.H., Hecht, F. and Motulsky, A.G.: Human embryonic hemoglobins. *Cold Spring Harbor Symposia on Quantitative Biology* 29, 327–31 (1964). Copyright (1964) by the Cold Spring Harbor Laboratory.

there is a single type of haemoglobin, with the composition $\alpha_2\epsilon_2$, then the main foetal form is produced (HbF with the structure $\alpha_2\gamma_2$) and in the adult there is the predominant HbA with the composition $\alpha_2\beta_2$, together with a minor component of the erythrocytes, HbA$_2$, with a structure of $\alpha_2\delta_2$. The α-chain of globin is involved in all of these types, but its partner changes during the development of the individual. The various globin chains are specified by different genes, and so the changing composition of haemoglobin is a manifestation of the activity of different genes. This change could be brought about within a single cell by a change in gene expression during the differentiation of the erythrocyte and this hypothesis is supported by the observation that a single cell may contain both β- and γ-globin chains. On the other hand there is evidence that the change of haemoglobin type may occur as a result of the proliferation of a different population of erythrocytes and their stem cells. In the frog *Rana catesbeiana* the new adult erythrocytes differ somewhat in shape from those of the tadpole. This new population of erythrocytes can be elicited prematurely in the life history of the frog by treatment with thyroxin [90].

There is clear evidence of some sort of feedback regulation in the synthesis of the globin chains since if the formation of one type of non-α chain is deficient, for instance in the case of genetic abnormality, then another non-α type of chain may be produced in quantity so as to compensate in part for the deficiency. For example, when there are defects in β-chain synthesis the level of γ-chains in the blood may remain high after birth, as foetal haemoglobin, HbF, persists into adult life.

79

The lens of the eye has proved a valuable system in which to study cell differentiation since the organ consists of a population of cells derived from an epithelium without the admixture of different types of cells from other sources. The value of the system and some findings derived from its study are set out in an article by Clayton [24]. The embryology of the eye has been studied for many years. The lens is derived in development from the ectoderm in response to an inducing stimulus from the developing optic cup (see Chapter 5). At first a vesicle is formed composed of a layer of epithelial cells which are similar to each other. Then some of these cells begin to change, beginning to elongate so that they eventually fill up the lumen of the vesicle. These elongated cells are the **primary lens fibres**. In the next stage of development a germinal zone of epithelial cells around the equator of the lens begins a process of division to form cells which are displaced posteriorly and then begin to differentiate and form more fibres, the **secondary lens fibres**. These secondary fibres are produced continuously throughout life and they form successive layers of cells around the primary fibres. However, apart from their origin and position, the secondary fibres are very similar to the primary fibres (Fig. 38).

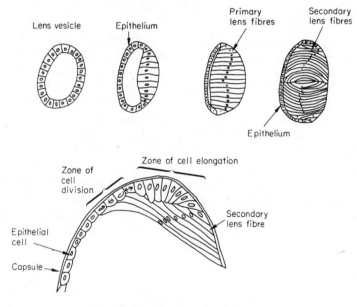

FIG. 38. The Development of the Lens

The normal differentiation of the fibre cell can be paralleled by a differentiation which has been achieved by cells under conditions of culture by Okada, Eguchi and Takeichi [95]. Dissociated cells from the chick lens epithelium were maintained in culture for periods of up to 50 days. Some of these cells differentiated and were similar to normal lens fibres as judged both by electron microscopy and by the accumulation of the lens-specific protein, δ-crystallin.

The general processes of development and the biochemistry of the fibre cells are very similar in all species of vertebrates, but much of the work on embryonic development and differentiation has been carried out with the chick and it is that species which will be mainly described here.

As the fibre cell differentiates from the epithelial cell there are a number of remarkable changes in its morphology and ultrastructure. A notable feature of the differentiation is the marked elongation of the cells and this is accompanied by an alignment of longitudinally oriented microtubules along the inside of the cell membrane [106]. At first the nucleus and nucleolus enlarge, both RNA and DNA synthesis occur, and the population of ribosomes within the cell increases. Then as the cell elongates it accumulates a very high concentration of a group of structural proteins, the **crystallins**, until eventually these proteins make up 30 per cent of the wet weight of the cell. At the same time there is a gradual loss of many of the cell organelles. The mitochondria are lost and the energy production of the cell comes to depend largely on anaerobic glycolysis. The ribosomes which are in the cell are found in smaller aggregates and eventually the cell nucleus disintegrates and the DNA is broken down. Thus the mature lens fibre cell, like the mammalian erythrocyte, is enucleate and represents the extreme case of terminal differentiation since it is clearly incapable of giving rise to any other type of cell.

Pearce and Zwaan [102] point out that the cellular elongation of the lens fibres is intrinsic to the cells and not the result of external forces, since the cells can elongate in culture. These authors believe that though the microtubules may be required for the change in cell shape, they are not alone responsible for the elongation.

The epithelial cells of the lens, besides giving rise to the cells which differentiate into fibres, also secrete the surrounding non-cellular capsule which consists of a protein akin to collagen and also of mucopolysaccharide material. During the development of some of the cells in the epithelium there is a switch in activity from the production of the capsule components to that pathway of development leading eventually to the lens fibre [24].

Lens protein

The crystallins which make up such a large part of the lens form three groups in the chick, known as α-, β- and δ-crystallins, and each of these groups consists of molecules which are themselves built up from a number of different types of subunit [141]. The number of different polypeptide chains which comprise the major structural proteins of the lens is about twenty plus several minor components. Not all of these polypeptides are synthesized in the early stages of lens development, but there is a gradual build up in the number of different polypeptides in the lens, and in the chick the composition of the lens proteins is still changing rapidly after hatching [140] (Figs. 39 and 40).

During lens fibre differentiation there are changes in the enzymes of the cells in addition to the changes in the structural proteins. The chief energy supply of the lens is derived from the glycolytic metabolism leading to the production of lactate. The pentose phosphate pathway accounts for only about 10 per cent of glucose metabolism and the poor oxygen supply to the lens resulting from the absence of a blood circulatory system prevents oxidative phosphorylation being carried out in most lens cells. This is also associated with the absence of mitochondria from the differentiated fibres. During differentiation there is a change in the types of lactate dehydrogenase isozymes present in the cells from a predominance of LDH-5 to an

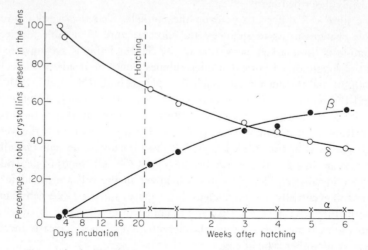

FIG. 39. Changes in the Protein Composition of the Chick Lens During Development
The composition of the lens as shown by quantitative immunoelectrophoresis.
From Truman, D.E.S., Brown, A.G. and Campbell, J.C.: The relationship between the ontogeny of antigens and of the polypeptide chains of the crystallins during lens development. *Experimental Eye Research* 13, 58–69 (1972).

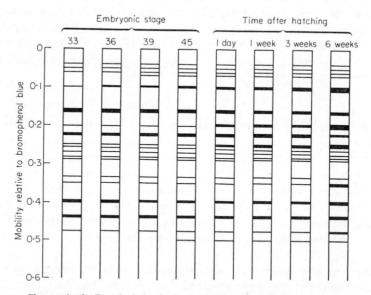

FIG. 40. Changes in the Protein Subunits in the Chick Lens during Development
Diagram of the results of electrophoresis of lens extracts on polyacrylamide gel in the presence of 6M-urea.
From Truman, D.E.S., Brown, A.G. and Campbell, J.C.: The relationship between the ontogeny of antigens and of the polypeptide chains of the crystallins during lens development. *Experimental Eye Research* 13, 58–69 (1972).

abundance of LDH-1 [97]. Nevertheless, the most conspicuous biochemical event in lens differentiation is the accumulation of the twenty or so polypeptides that make up the bulk of the crystallins.

Control of crystallin synthesis
It was observed by Scott and Bell in 1964 [119] that in early embryonic cells, and also in the developing lens at an early age, the inhibition of RNA synthesis by actinomycin D was soon followed by a reduction in the rate of protein synthesis. However, in older lenses the sensitivity of protein synthesis to actinomycin D diminished, and though the synthesis of RNA was still inhibited protein synthesis appeared to continue in the presence of the inhibitor. Now it is generally supposed that protein synthesis is affected by actinomycin D because mRNA synthesis is first affected and as the messengers of the cell are broken down, so protein production is diminished. It appeared from the work of Bell and his colleagues that in the lens the mRNA molecules became more stable as the lens developed. Studies of the effect of actinomycin D on the structure of polysomes in the lens at different stages of development confirmed this view of the stability of the mRNA. A more detailed study showed that the sensitivity of protein synthesis to actinomycin D varied from one polypeptide to another, and it appears that the stability of the mRNA molecules for different proteins may vary quite considerably even within the same type of cell. This implies that the rate of breakdown of the mRNA is not simply dependent upon the amount of ribonucleases within the cell but is a property of the mRNA itself [24].

As a method of measuring the stability of mRNA molecules actinomycin D is a rather indirect tool, and it also appears to have other effects than the well-known ones upon transcription. In the case of the lens a more direct method has been applied to estimating the rate of decay of mRNA based on examining the turnover of RNA associated with fractions of polysomes selected immunologically according to the polypeptides still attached to them. This method has given confirmation that as lens development continues in the embryo there is a stabilization of mRNA for the crystallins, but that this effect varies between the mRNA molecules coding for different polypeptides. Furthermore, the mRNA involved in the synthesis of the enzymes of the lens appears to be less stable than that specifying the crystallins [25].

The lens has been successfully used in attempts to isolate functioning mRNA. As in the reticulocyte, the mRNA is apparently present in the form of a complex with protein, as a messenger ribonucleoprotein particle (mRNP particle) [158]. Two main classes of mRNP particles have so far been resolved in both chick and bovine lenses, but the number of different types of mRNA present must be much greater than this. Since changes in the stability of mRNA are apparently not a property of the cell but rather of the specificity of the messenger molecules it is a problem to understand how the stabilization may be brought about during development: protein associated with the mRNA in the mRNP particle might conceivably have some function in determining the rate of breakdown of the messenger.

In describing the differentiation of the lens fibre emphasis has been placed on the translational control of protein synthesis, since it appears that in the lens the accumulation of large amounts of a limited range of proteins is facilitated by increasing the duration of the active life of the messenger molecules. There is evidence of similar stabilization in some other differentiated cells, such as those of the skin, as well as in the reticulocyte described above. The lens provides a good example of the role that control of translational processes may have during cytodifferentiation.

Lens regeneration

Mention was made in Chapter 4 of the way in which various eye tissues could dedifferentiate and then redifferentiate to form lens tissue if the original lens is removed from the eye of various animals. In the larvae of the toad *Xenopus laevis*, for instance, either iris, retina or cornea cells can give rise to lens cells after dedifferentiation. Indeed, if portions of the cornea of the eye are maintained in culture in the absence of a lens some of the cells divide to form an outgrowth of the cornea and may then change their morphology and histological staining properties until they resemble those of lens fibres. Moreover, the antigens which are characteristic of the differentiated lens can then be detected. The addition of portion of lens to these cultures of cornea when they are set up appears to inhibit the differentiation. It seems that it is the presence of a lens which normally prevents the cornea from giving rise to lens tissue, indicating the existence of some sort of feedback control of this differentiation [22].

LIVER

The differentiation of lens fibres and of erythrocytes involves the reduction of the activities of these cells to the production of large amounts of a limited number of different types of polypeptide chains and this is accompanied by a drastic simplification of the organization of the cell. The differentiation of liver, on the other hand, is accompanied by a very considerable increase in the biochemical repertoire of the cell, including the synthesis of a very wide range of enzymes, while at the same time the ultrastructural organization of the cells remains complex. In fact, a major difficulty in describing liver differentiation is the problem of systematizing the great range of changes that take place as the organ develops.

The liver originates in the embryo as an outgrowth of the gut and associated mesenchyme and the formation of the tissue involves the differentiation of a number of different types of cell, though the **hepatocytes** make up the bulk of the organ. The functions of the liver include transformations of a wide range of substances, breakdown and interconversions of amino acids, and conversion of surplus amino compounds into urea; synthesis, storage and breakdown of glycogen, leading to maintenance of blood glucose levels; the synthesis of lipids, and the production of important blood plasma proteins. Not all of the enzymes which are characteristic of adult liver appear at the same time, nor is there a steady rate of accumulation of different enzyme activities, but rather there are certain critical times at which whole groups of enzymes make their appearance. In mammals, such as the rat, the most active times of appearance of new enzyme activities are in the late foetus, then at about the time of birth, and at the time of weaning. We can understand the timing of the formation of some of these enzymes in terms of their relation with the biology of the whole animal. For example, when the young rat is weaned from a diet of milk on to a more varied diet its food is richer in carbohydrates and poorer in fats and protein. The increase in the activity of the enzymes needed for the synthesis of lipids and the interconversion of amino acids which takes place at the time of weaning can be seen as adaptations to the dietary change.

The activities of different enzymes in the developing liver are discussed in detail in an article by Moog [88] and are well summarized by Greengard [46], so that only a few cases will be considered here. Greengard has studied enzyme differentiation in rat liver and has tabulated many of the enzymes of the liver in groups or 'clusters' according to the time at which their activity increases. Uridine diphosphate glucose glycogen glycosyl transferase (E.C.2.4.1.11) is characteristic of the late foetal cluster. As is shown in Fig. 41, it rises rapidly in activity about 3 days before birth.

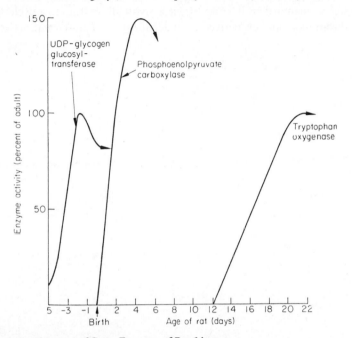

FIG. 41. Development of Some Enzymes of Rat Liver
    After Greengard, O.: Enzymic differentiation in mammalian liver. *Essays in Biochemistry* 7, 159–205 (1971). Copyright (1971) by the Biochemical Society.

It is responsible for the polymerization of glucose units to form glycogen and its activity leads to the formation of a store of liver glycogen before birth. This glycogen provides a reserve of energy over the time of birth when the placenta no longer nourishes the foetus and while feeding by sucking is still being developed.

While placental nutrition is constant in the foetus, the life of the young mammal is dependent upon intermittent feeding and the concentration of glucose in the blood is maintained at a constant level by drawing on reserves of glycogen in the liver and by conversion of amino acids to glucose from time to time. An essential and apparently rate-limiting step in such glucose synthesis is the conversion of oxaloacetate to phosphoenolpyruvate, so that citric acid cycle intermediates can be brought back into the Embden-Meyerhof pathway. This critical step is catalysed by phosphoenolpyruvate carboxylase (E.C.4.1.1.32) which rises greatly in activity at about the time of birth (Fig. 41).

The third enzyme shown in Fig. 41, tryptophan oxygenase (E.C.1.13.1.12), probably serves to remove any dietary tryptophan which is in excess of bodily needs. This situation arises in the rat at the time of weaning.

At certain times of development whole new pathways of metabolism become active and these may be associated with the almost simultaneous appearance of several enzymes. A clear case in the liver is seen in the enzymes of the urea cycle. The enzymes carbamylphosphate synthetase (E.C.2.7.2.5), ornithine carbamoyl-transferase (E.C.2.1.3.3), argininosuccinate synthetase (E.C.6.3.4.5), arginosucci-nate lyase (E.C.4.3.2.1) and arginase (E.C.3.5.3.1) are all members of the late foetal cluster in mammalian liver and their activity allows the urea cycle to convert into urea the ammonium ions derived from deamination of amino acids (Fig. 42) [46].

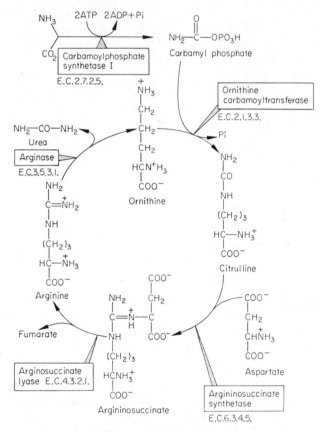

FIG. 42. The Urea Cycle

Regulation of liver enzymes

The increase in activity of certain enzymes at particular times might be regulated in a way which depended upon the age of the foetus, by some kind of inherent biological clock mechanism, or the changes might result from the interaction of the liver cells with their changing environment. With regard to many enzymes, the experimental evidence supports this latter view since if foetal rats are delivered prematurely as a result of surgery then they show a premature rise of many enzyme activities. One possible stimulus for these changes is the lowering of blood sugar concentrations (hypoglycaemia) that occurs at birth. Hypoglycaemia stimulates the pancreas to secrete glucagon, and this hormone has been shown to have a marked effect on several liver enzymes. Injection of glucagon into foetal rats can induce rises in activity of several enzymes that usually become more active at birth, for example phosphoenolpyruvate carboxylase (E.C.4.1.1.32), tyrosine aminotransferase (E.C.2.6.1.5), L-serine dehydratase (E.C.4.2.1.13) and glucose 6-phosphatase (E.C.3.1.3.9)[46]. In some instances it can be clearly shown that the rises in enzyme activity are the result of the synthesis of new enzyme molecules and not merely the activation of pre-existing molecules. Sometimes a single enzyme may be influenced by more than one hormonal stimulus as happens with tyrosine aminotransferase (Table 11).

TABLE 11. Susceptibility of Liver Enzymes to Hormonal Regulation

	Glucagon	Hydrocortisone	Thyroxin
Phosphoenolypyruvate carboxylase E.C.4.1.1.32	+		
Tyrosine aminotransferase E.C.2.6.1.5	+		
L-Serine dehydratase E.C.4.2.1.13	+		
Glucose 6-phosphatase E.C.3.1.3.9	+		+
UDPG-glycogen glycosyl transferase E.C.2.4.1.11		+	
Arginase E.C.3.5.3.1		+	+
NADPH–cytochrome c dehydrogenase E.C.1.6.99.2			+
Ornithine aminotransferase E.C.2.6.1.13		+	
Malate dehydrogenase (NADP) E.C.1.1.1.40			+
Tryptophan oxygenase E.C.1.13.1.12		+	
Pyruvate kinase E.C.2.7.1.40			+

If glucagon is injected into very early foetuses there is no response in terms of enzyme activity. Tyrosine aminotransferase (TAT) is stimulated by glucagon on the 20th day of gestation in the rat, but not on the 18th day. Before this time the liver is not apparently competent to respond. We thus have a situation analogous to the phenomenon of competence in embryonic induction (see Chapter 5). In this case of liver synthesis of TAT we can give a more detailed explanation of the phenomenon. It is believed that like many other hormones glucagon produces its effects via cyclic adenosine-3'-5'-monophosphate (cAMP). If cAMP is administered to the foetus TAT can be produced on the 18th day of gestation. The change which apparently takes place after 19 days of gestation is the ability of liver to produce cAMP in response to glucagon. The foetal rat does not itself develop the ability to produce glucagon in response to hypoglycaemia until after 22 days of gestation. After birth the level of TAT in the liver becomes susceptible to further hormonal influences, some of which are indicated in Fig. 43. As the animal matures more and more stimuli are present

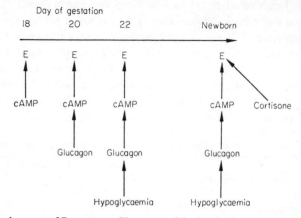

FIG. 43. Development of Response to Hormones of the Synthesis of Tyrosine Aminotransferase in Foetal Rat Liver
From Greengard, O.: Enzymic differentiation in mammalian liver. *Essays in Biochemistry* 7, 159–205 (1971). Copyright (1971) by the Biochemical Society.

which tend to stimulate TAT activity, so that the removal of any one of them will not neccessarily reduce its concentration. It may well be difficult to remove all of the influences which promote enzyme synthesis in the fully differentiated liver. It is possible that the stability of cellular differentiation may frequently be due to the presence of a wide range of stimuli acting from outside the tissue.

The control mechanisms of the synthesis of TAT have been studied in some detail by Tomkins and his colleagues [137], making use of a strain of liver cells that can be maintained in culture, rat hepatoma cells (HTC cells). TAT can be induced in these cells by the action of a synthetic adrenal steroid, dexamethasone phosphate. About 10 hours after treatment with the hormone the enzyme rises in activity to between 5 and 15 times that before hormone treatment, and by immunological assay it can be shown that the rise in activity is due to the synthesis of new enzyme molecules. There is no corresponding increase in the rate of synthesis of the general cell proteins.

Using cultures of HTC cells in which cell division was synchronized it was shown that though TAT could be synthesized at all stages of the cell cycle, the rate of synthesis could only be influenced by the hormone during a limited period and that for a period of several hours on either side of mitosis the enzyme was not inducible.

When the steroid hormone was removed from the HTC cells there was a cessation of the synthesis of tyrosine aminotransferase and a gradual decline in its activity. Inhibitors of RNA synthesis, such as actinomycin D, inhibited the induction of the enzyme as might be expected. However, it was found that after the enzyme was induced inhibition of RNA synthesis caused the rate of synthesis of the enzyme to remain high even after the inducer had been removed, suggesting that RNA synthesis is necessary to inhibit the synthesis of TAT in the absence of the inducer. Indeed when actinomycin D is added after the removal of the inducer the level of the enzyme can rise again, a phenomenon that Tomkins refers to as **messenger rescue** (Fig. 44). These findings are rather remarkable and have led Tomkins to propose a hypothetical scheme of regulation of tyrosine aminotransferase synthesis which involves both a

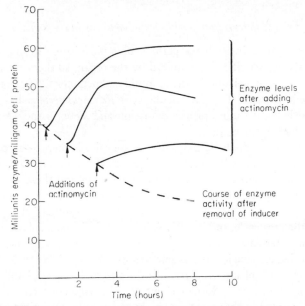

FIG. 44 Effect of Actinomycin D on Tyrosine Aminotransferase Synthesis in Cultured Liver Cells

From Tomkins, G.M., Gelhrter, T.D., Granner, D., Martin, D., Samuels, H.H. and Thompson, E.B.: Control of specific gene expression in higher organisms. *Science* **166**, 1474–1480 (1969). Copyright (1969) by the American Association for the Advancement of Science.

structural gene for the enzyme and also a regulator gene. The structural gene produces the messenger RNA (TAT-mRNA) which when translated gives rise to the enzyme, while the regulator gene produces a messenger which gives rise to a repressor substance. The repressor substance not only inhibits the translation of the TAT-mRNA but actually stimulates its breakdown. The hormone is presumed to act by combining with the repressor and inactivating it. In this way the TAT-mRNA becomes stabilized, accumulates and can be transcribed. When RNA synthesis is inhibited after induction and when the level of TAT-mRNA is high, the repressor cannot be synthesized and the TAT-mRNA can then be translated and is not broken down. This explains how the actinomycin D can lead to the retention of tyrosine aminotransferase synthesis even after the inducer is removed, and can indeed stimulate synthesis of the enzyme when there would otherwise be a steady decline following the removal of the inducer, as in the messenger rescue experiments.

In the non-inducible phases of the cell cycle it is supposed that neither the structural gene nor the regulatory gene is transcribed, though any TAT-mRNA which may remain can continue to be translated.

The rather complex experimental data about the regulation of tyrosine aminotransferase synthesis and the detailed hypothesis of Tomkins and his group have been described at some length since this is one of the most clearly established instances of the role of regulatory genes in higher organisms, though it is not the only one. This system of enzyme regulation continues to provide a basis for experimentation on translational control in differentiated cells [138].

Liver differentiation during metamorphosis in amphibians

Just as the mammalian liver shows periods of rapid change during development, when new enzymatic activities are acquired by the organ, so there are times of rapid change in other vertebrates, and in the amphibians metamorphosis from the tadpole to adult is one such time. The liver of the frog *Rana catesbeiana* has proved to be valuable in biochemical studies of differentiation and some of the work on this has been reviewed by P.P.Cohen [27].

As the liver of the tadpole develops at metamorphosis there is no cell division in the organ and so the changes that occur represent modifications of the activity of existing cells. The increases in the activity of many enzymes and those of the urea cycle are produced in a manner that is similar to that of mammalian liver (see p. 86). One particular enzyme, carbamoylphosphate synthetase I (CP-synthetase) is rate-limiting in the activity of the urea cycle, and its activity increases in metamorphosis slightly earlier than that of the other enzymes of the cycle (Fig. 45). By use of an antibody prepared against the purified enzyme it was possible to show that the increase in activity of the enzyme was the result of the synthesis of new enzyme molecules and was not produced by activation of a pre-existing enzyme. The enzyme appeared in all cells of the organ and was located in the mitochondria in these cells. Investigation of the regulation of the enzyme is facilitated by the fact that CP-synthetase I, along with many other liver enzymes, could be induced to appear prematurely by treatment of the tadpoles with the hormone thyroxine.

Thyroxine treatment of the tadpoles resulted in an increase in the rate of synthesis of rRNA and tRNA and also of a fraction of RNA which is synthesized rapidly and has a base composition corresponding to that of the DNA. This fraction of RNA has been supposed to represent mRNA. The rise in rate of RNA synthesis precedes the changes in CP-synthetase I activity and the gross morphological changes of metamorphosis. Studies of the chromatin isolated from the liver of the tadpoles showed that the ability of the chromatin to act as a template for RNA synthesis when RNA polymerase was added increased following administration of thyroxin to the tadpoles. However, when purified DNA was tested as a template for RNA polymerase activity it was found that it was unaffected by thyroxin treatment of the tadpoles. It appears then that one of the changes brought about by thyroxin is the unmasking of portions of the DNA in the chromatin so that it can act as a template for the production of RNA. It seems likely that some of this RNA which is synthesized after hormone treatment may be the mRNA for liver enzymes, including the rate-limiting carbamoylphosphate synthetase.

It has been possible here to mention only a very few of the biochemical changes that take place as liver cells differentiate. It is clear even from these that this differentiation is a gradual process and that the tissue may be clearly seen to be liver and yet not have all of the biochemical characteristics of adult liver. Many of these characteristics follow from the induction of specific enzymes by hormones, while other changes in the enzymatic repertoire of the liver can be brought about by alterations in the physiological or nutritional status of the animal. It is not at all easy to draw a distinction between those enzymes which are fundamental to the differentiation of the liver and those which represent very short-term modulations in the activity of the

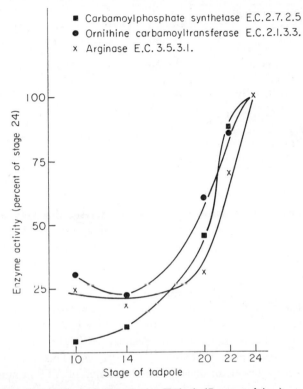

FIG. 45. Urea Cycle Enzymes in Developing Tadpole (*Rana catesbeiana*)
From Cohen, P.P.: Biochemical differentiation during amphibian metamorphosis. *Science* **168**, 533–43 (1970). Copyright (1970) by the American Association for the Advancement of Science.

organ. Indeed, from what we know of the biochemistry of this organ it seems that differentiation and modulation do not represent distinct processes but are merely the extreme ends of a spectrum of changes that can occur in this organ.

SKELETAL MUSCLE

The differentiated muscle fibre is not a single cell but is a **syncytium** containing many nuclei and derived from the fusion of hundreds of cells, the **myoblasts**. Once the cells have fused there is no further DNA synthesis or division of the nuclei and the fibre can grow only by the addition of more mononucleate myoblasts.

The chief function of muscle is, of course, contraction and there are many proteins which are characteristic of muscle and which are associated with contractility: actin and myosin, tropomyosin, troponin and actinin (see Table 12). These proteins have

TABLE 12. Proteins Characteristic of Muscle

	Molecular weight	Per cent of total myofibril protein
Structural Proteins		
Myosin	500 000	50–60
Actin	45 000	20
Tropomyosin	70 000	7–10
Troponin	80 000	2–5
α-Actinin 6S	160 000	
α-Actinin 25S	3 200 000	2–10
β-Actinin	130 000	1–2
Enzymes		
Creatine kinase (E.C.2.7.3.2)		
Phosphorylase a (E.C.2.4.1.2)		

long been regarded as restricted to muscle, but it has recently been found, as a result of the development of very sensitive methods of analysis, that very similar, if not identical proteins, may occur in other cells, particularly associated with contractile activity. It seems that the microfilaments of cells, which are probably involved in cell movement and changes in cell shape, are composed of proteins closely resembling actin [104].

The energy for the contraction of muscle is derived from the hydrolysis of adenosine triphosphate (ATP), and a store of chemical potential energy is maintained in muscle in vertebrates in the form of the compound creatine phosphate. This serves to regenerate ATP from adenosine diphosphate (ADP) by the reaction catalysed by the enzyme creatine phosphokinase (E.C.2.7.3.2). This enzyme is another protein which is characteristic of muscle. The formation of ATP in muscle involves chiefly the glycolytic pathway, the citric acid cycle and oxidative phosphorylation and the enzymes involved in these processes, though present in almost all tissues, are particularly concentrated in muscle (see Chapter 3).

The ultrastructural organization of muscle tissue is very distinct and also somewhat complex. Details of this ultrastructure and of the arrangement of the molecules of actin and myosin which is believed to give rise to the structure, will be found in most textbooks of cell biology. The development of this ultrastructure represents the end-product of muscle cell differentiation and provides a microscopic criterion of that differentiation. However, biochemical studies provide a more sensitive test since it is possible to detect the production of some muscle enzymes or the accumulation of the major structural proteins some time before striations in the fibres are detectable microscopically (Fig. 46).

The production of the muscle fibre can be regarded conveniently as having two stages: (i) the origin and multiplication of the mononucleate myoblast and (ii) the further differentiation to produce the fibre after cell fusion. By manipulation of tissue culture procedures it is possible to separate these two phases and to initiate fusion of cells in a controlled manner. Important work in this field has been carried out by Konigsberg [104].

Embryonic muscle tissue can be dissociated into separate cells by mincing and then treating briefly with trypsin to produce mainly mononucleated myoblasts. These cells can be maintained in culture and will then divide quite rapidly after attaching

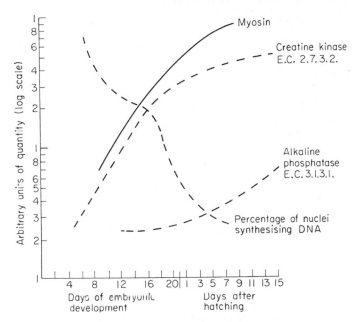

FIG. 46. Enzyme Changes in Developing Chick Skeletal Muscle
    After Hauschka, S.D.: Clonal aspects of muscle development and the stability of the differentiated state. *The Stability of the Differentiated State*, ed. by Ursprung, H. (1968). Berlin-Heidelberg-New York: Springer.

themselves to the bottom of the culture dish. If they are allowed to multiply until they form a continuous layer on the bottom of the dish, when they are said to have grown to confluence, they then begin to aggregate to form multinucleate muscle fibres. Following the cell fusion some of the enzymes characteristic of muscle increase rapidly in amount, such as creatine kinase (E.C.2.7.3.2) and adenylate kinase (E.C.2.7.4.3) (Fig. 47). That these increases in activity represent the production of new proteins is suggested by the prevention of their appearance following the addition of inhibitors of protein synthesis.

If individual myoblasts are taken before they have grown to confluence and are transferred to another culture dish they will continue to divide, and the progeny of these cells will retain the capability of fusing to form fibres when the cell density becomes sufficient. It was found that all the progeny of a single cell remain myoblasts or fuse to form muscle fibres. By repeatedly transferring cells to new culture dishes before they grew to confluence it was possible to maintain a population of cells for many cell generations for more than two years in culture. These cells, though not fully differentiated, were yet capable of specific differentiation if they were allowed to grow to confluence. Moreover, at no time did they show any tendency to differentiate in any other manner than into muscle fibres [160].

As myoblasts *in vivo* approach the time at which they fuse there is a reduction in the rate of synthesis of DNA which is accompanied by a loss of activity of the enzyme DNA polymerase. There is no reason, however, to suppose that the fusion of the cells and the ending of DNA synthesis are causally related. Nevertheless,

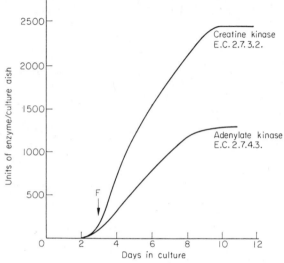

FIG. 47. Changes in Enzyme Activity in Cultured Muscle Cells
   F represents time of fusion of myoblasts
   From Yaffe, D.: Cellular aspects of muscle differentiation *in vitro*. *Current Topics in Developmental Biology* 4, 37–77 (1969). Copyright (1969) Academic Press, New York.

muscle tissue does illustrate the view, which has been widely held, that there is an antagonism between cell division and the synthesis of large amounts of tissue-specific proteins. It is possible that there is a competition for energy supply between the processes of cell division and those of protein synthesis and that the production of large amounts of actin and myosin only becomes possible when cell division ceases.

MAMMARY GLAND

The mammary gland provides us with another system in which we can investigate the stimuli bringing about a specific differentiation, since the hormones present during pregnancy can be administered in culture to mammary tissues taken from virgin mice [143]. The mammary gland differentiates in two stages: the system of **ducts** is developed by the invagination of the germinal layer of the epidermis during adolescence and the first half of pregnancy, while the **alveolar lobules**, which synthesize and secrete the characteristic substances of milk, are formed by the differentiation of epithelial cells during the second half of pregnancy. The specific substances which these alveolar cells secrete include the proteins casein, α-lactalbumin, β-lactoglobulin and lactose synthetase.

94

The ductal tissue can develop in culture in a synthetic medium devoid of hormones but the alveolar structures will only differentiate in a medium to which the hormones insulin, hydrocortisone and prolactin have been added. The differentiation can be followed microscopically by the enlargement of the epithelial cells, the prominence of their nucleoli and the accumulation of fat droplets in the cells. At the ultrastructural level there is an increase in endoplasmic reticulum and ribosomes within the cells as they differentiate, with dilation of the Golgi apparatus and an increase in the number of microvilli on the free surface of the epithelial cells. A biochemical index of differentiation is provided by the synthesis of the specific proteins such as α-lactalbumin or casein. The latter is a phosphoprotein and it can be detected by the incorporation of radioactive phosphate into the total proteins of the cells. This simple index of differentiation has proved of great value in studies of the effect of various hormones on the specific protein synthesis of differentiating mammary gland.

While the addition of insulin to mammary gland tissue in culture helps to maintain a low level of casein synthesis and has some slight stimulatory effect on protein synthesis in general in the cells, the hormones prolactin and the hydrocortisone have specific effects upon the synthesis of milk proteins and do not have any effect on the general rate of protein synthesis in the system. A very notable feature of the effect of hormones on this gland is the **synergistic** effect of insulin, prolactin and hydro-cortisone: that is, their combined effect is greater than the sum of their individual effects. Insulin seems to stimulate mitosis in the epithelial cells, and the effect of the other hormones is apparent only when they are acting on dividing cells. Thus, addition of insulin increases the effectiveness of the other hormones (Fig. 48) [143].

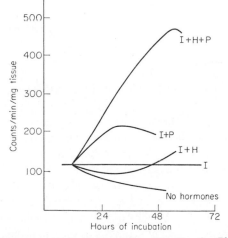

FIG. 48. Effect of Hormones on the Incorporation of Radioactive Phosphate into Casein by Mouse Mammary Gland Tissue in Culture
    I=insulin; H=hydrocortisone; P=prolactin
    After Turkington, R.W.: Hormone-dependent differentiation of mammary gland *in vitro. Current Topics in Developmental Biology* 3, 199–218 (1968). Copyright (1968) Academic Press, New York.

The cells of the epithelium which secrete the specific proteins of milk are evidently formed by a mitotic division which must occur in the presence of hydrocortisone and prolactin. This is one of a number of examples in cell differentiation where a mitotic cell division is a turning point in the development of the line of cells. One possible explanation of such a requirement for division if differentiation is to take place could be that the sites of regulation of the genes may become accessible to the specific stimulus only at a time when the nuclear membrane has broken down as it does in mitosis.

PANCREAS

Differentiation of the pancreas involves the formation of a number of different cell types, but most studies so far have considered the organ as a whole. The **exocrine** portion of the pancreas secretes enzymes via a system of ducts and these enzymes are synthesized in the **acinar** cells. Some of the enzymes are produced in an inactive form, or **zymogen**, which is activated by the breaking of peptide bonds to give the functioning form of the enzyme. Thus trypsinogen is activated to form trypsin.

The other portion of the pancreas is the endocrine region, composed of the Islets of Langerhans, and here there are two types of cell, one type secreting insulin and

TABLE 13. Proteins Characteristic of the Pancreas

Type of cell		
Islet cells:	A cells	Glucagon
	B cells	Insulin
Acinar cells		Amylase
		Chymotrypsinogen
		Trypsinogen
		Procarboxypeptidase a
		Procarboxypeptidase b
		Lipase
		Ribonuclease

the other type producing glucagon. Indeed insulin itself is produced from a precursor molecule, proinsulin, which is activated by cleavage of the polypeptide chain [23].

The pancreas is one of the organs which is produced by an inductive interaction between an epithelium and the surrounding mesenchyme and this process has been extensively investigated by Grobstein's transfilter technique as outlined in Chapter 5. The epithelium concerned is an outgrowth of the gut and at first it shows a very rapid cell division. During this stage several of the specific pancreatic proteins can be detected at a very low concentration. Later on, for example after about 14 days gestation in the mouse, division of the cells in the central acinar region stops and there is a very great increase in the rate of synthesis of specific proteins (Fig. 49).

Wessels and Wilt [155] studied the effect of actinomycin D on the formation of the tissue-specific proteins of the pancreas. They found that if they treated the organ with the inhibitor at the stage when cell division had stopped in the central region they could inhibit the synthesis of all classes of RNA by 70 per cent in all regions of the embryonic organ. Nevertheless there was no inhibitory effect on the formation of amylase and zymogen granules. If, however, actinomycin D was applied before the cell division in the central region had ceased, then the specific pancreas functions did not develop. They concluded from these experiments that the mRNA molecules which specify the specific pancreatic proteins are synthesized during the stage of rapid cell division, but that they are only extensively translated after the end of cell division. We see in this work a further illustration that the presence of certain mRNA

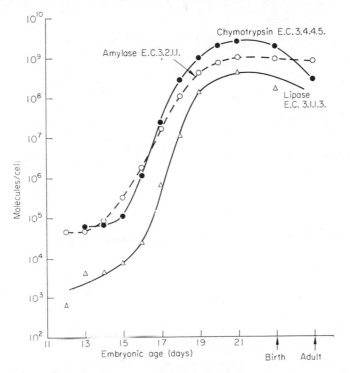

FIG. 49. Ontogeny of Some Mouse Pancreatic Enzymes

After Rutter, W.J., Kemp, J.D., Bradshaw, W.S., Clark, W.R., Ronzio, R.A. and Sanders, T.G.: Regulation of specific protein synthesis in cytodifferentiation. *Journal of Cell Physiology* **72**, supplement 1, 1–18 (1968). Copyright (1968) The Wistar Press.

molecules in the cell is not sufficient to bring about the synthesis of the protein which they specify, but that there is evidently some regulating influence acting at the stage of translation.

The differentiating pancreas brings us another example to consider in view of the antagonism which apparently exists between cell division and differentiation. Here the rate of formation of specific substances is only slow while division is rapid and then specific synthesis increases as mitosis slows down. The two processes are clearly not mutually exclusive though they may be somewhat antagonistic.

7

SOME CONCLUSIONS

The first five chapters of this book have discussed some of the general features of cell differentiation in an attempt to illustrate the key points where we must seek an understanding of the process, and Chapter 6 has given us examples which we must try to keep in mind while coming to a conclusion. It must be stressed that such a conclusion can only be of an interim nature since the phenomenon of cytodifferentiation still poses many more questions than it answers, and current research is continually providing us with more data which we must accommodate in our view of the subject. Some conclusions we may reach are outlined below and then certain of these are enlarged upon later.

1 The regulation of protein synthesis is a central problem to an understanding of cell differentiation.

2 Protein synthesis can be regulated in a variety of ways, which can be conveniently classified into transcriptional regulation and translational regulation.

3 Translational control mechanisms may themselves be the result of the regulation of the activity of other genes.

4 Selective activity of genes occurs during cell differentiation and provides an explanation of much of the regulation of protein synthesis.

5 The regulation of gene activity includes processes acting on short time scales of the order of minutes, as well as regulatory processes with much longer time scales, of the order of years.

6 At least some aspects of cellular regulation must be inherited over several cell generations to account for the stability of differentiation.

7 Some, at least, of the genes are regulated by signals from the cytoplasm, or mediated via the cytoplasm.

Let us now consider some of these arguments in further detail.

1 In Chapter 2 a variety of examples were discussed to illustrate the role of enzymes in the direction of the metabolism of the cell, while many of the examples discussed in Chapter 6 have shown that cell differentiation involves the synthesis of specific enzymes and structural proteins or is accompanied by the development of specific patterns of enzyme concentration. The production of other tissue-specific

substances such as polysaccharides, or small molecules, such as thyroxine by the thyroid or apatite in bone, are the consequence of the existence of specific enzyme systems. The regulation of the enzymes in tissues covers not merely the presence or absence of the enzyme but in many cases involves a determination of the amount of enzyme produced.

2   The multitude of steps in the process of synthesis of a complete protein molecule were outlined in Chapter 3. It has proved convenient to distinguish the synthesis of the template RNA molecule under the direction of the base sequence in DNA as the process of transcription, while the many activities which utilize this template for the production of a protein molecule are grouped together as translation.

3   The participating molecules in protein synthesis, such as polymerizing enzymes, tRNA molecules, GTP and so on, are themselves, directly or indirectly, the products of genes and so are subject to regulation either at transcription or later stages. Ribosomal RNA molecules or tRNA molecules are the end-products derived from genes with relatively little processing, while amino acid activating enzymes could be regulated in concentration by any of the processes which regulate any other protein. If we found that a particular protein in a cell was not synthesized because of the absence of a particular activating enzyme we would say that the protein in question was subject to translational control, but the absence of the activating enzyme might itself be the result of either transcriptional or a translational control. This chain of cause and effect points to a crucial importance of gene transcription in the regulation of the synthesis of proteins.

Another process which might have a role to play in the regulation of differentiation is the processing of the RNA after it has been synthesized on the DNA template and before it leaves the nucleus. It has been suggested that there could be a selective breakdown of the newly synthesized RNA and that in different tissues different portions of this RNA might enter the cytoplasm to be translated. No convincing mechanism has yet been proposed of how this might be regulated, and we shall see that quite different functions of nuclear heterogeneous RNA have been postulated.

4   The conclusion that cellular differentiation is a consequence of the expression in different cells of different portions of the genome has become almost accepted as an axiom in developmental biology and our hypotheses of the mechanism of differentiation generally assume that this is so. If we can explain the phenomenon of cyto-differentiation in the majority of cases where the genome is not itself modified, then we should have little difficulty in accommodating the relatively few cases where the genome appears to alter. We may list some generalizations about the manner of differential gene activation:
(a) groups of functionally related genes are often activated together, even though they may not be genetically linked;
(b) in different tissues different patterns of genes are activated, some active genes being exclusive to one tissue, though many more are common to a wide range of tissues;
(c) a particular gene may become active under a number of different circumstances.

As yet the evidence of differential gene activity is largely indirect. Puffing patterns in polytene chromosomes give some evidence, but such cells are not typical, and much of the evidence put forward from hybridization studies of the RNA content of cells probably indicates differential synthesis of the repetitious portions of the genome rather than of the mRNA coding for enzymatic proteins. This evidence was reviewed in Chapter 5.

5 and 6   Some cells show very rapid responses to stimuli and may begin to synthesize specific products within an hour of the reception of a stimulus, while in other systems days may pass between the induction of differentiation and the first specific syntheses. Of course, the sensitivity of methods of analysis might account for the apparent differences in time scale in differentiation, or it might be that the specific syntheses being examined do not represent the first substances that are produced in response to the stimulus. Even so, it appears that there are very different time scales in operation in different types of cellular differentiation, and any general model proposed should account for these variations.

6   In some cases when a cell differentiates it ceases to be capable of cell division, as in the case of the lens fibre or the myoblast after fusion. On the other hand it frequently happens that a differentiated cell may be capable of division and that its progeny remain of the same cell type. Cultured fibroblasts can pass through many cell divisions and still retain their characteristic morphology and ability to secrete collagen. Moreover, many cell types which, though not terminally differentiated, are nevertheless determined upon a particular course of differentiation are capable of division and still remain determined in the same manner. An example of this is seen in the retinal pigment cell, described in Chapter 1. It is clear that the mechanism which brings about cellular differentiation must be capable of surviving cell division and must pass equally to the daughter cells after mitosis and cytokinesis.

7   While the survival of the differentiated state in cultured cells indicates that the cell may contain within itself the regulators of its own metabolism, it is clear that in many cases differentiation occurs after a stimulus has been applied to a cell or population of cells from outside and that this stimulus may act through the cytoplasm. The influence of the cytoplasm on the nucleus was discussed in Chapter 5 and there also some instances of hormones influencing the nucleus were discussed. Further examples were seen in the specific cellular systems described in Chapter 6. A number of hormones influence the pattern of protein synthesis in the liver and in the lactating mammary gland. Any generalized hypothesis of differentiation must also accommodate the regulatory role of hormones and also of embryonic inducing substances.

Some researches, such as those of Tomkins on the induction of tyrosine aminotransferase in the liver, have drawn attention to the importance of the cell cycle in responsiveness to external stimuli, and it has been proposed by Gurdon and Woodland [54] that one of the significant ways in which mitosis may play a part in differentiation is in providing a period in which the chromosomes are not enclosed within the nuclear membrane, so that substances present in the cytoplasm may be able to act upon the genetic material.

Higher organisms stand in marked contrast to bacteria and other prokaryotes in their ability to undergo cytodifferentiation and it might be expected that this ability is correlated with some of the distinctive features of the genetic apparatus of the eukaryotes. These features include the presence of chromosomal proteins, the large amount of DNA in the nucleus, especially that large proportion of DNA which appears from hybridization studies to have a repetitious base sequence, and the high proportion of RNA which is synthesized but which does not leave the nucleus, the nuclear heterogeneous RNA. An attempt to bring these observations into a unified theory of gene regulation in higher organisms was published by R.J.Britten and E.H.Davidson in 1969.

The theory of Britten and Davidson

In order to understand the theory proposed by Britten and Davidson [17] it is necessary to know the terminology that they use. Those sequences of DNA which specify the mRNA for the enzymes and structural proteins of the cell, and also those which code for other classes of RNA which enter the cytoplasm, viz. rRNA and tRNA, are called **producer genes**. It is proposed in the hypothesis that each of these genes is regulated by an adjacent sequence of DNA, the **receptor gene**. The precise mechanism of this regulation is not elaborated in the theory and forms no essential part of the model, but it could, for instance, be by the action of a product bringing about the removal of histone repressors from the structural gene.

The receptor genes are themselves activated by interaction with a species of RNA, called **activator RNA**, and again it is not necessary to specify the details of the mechanism involved. Activator RNA is the postulated product of sequences of DNA known as **integrator genes**. Whereas the receptor genes and producer genes are supposed to be closely linked, it is not necessary that the integrator genes should be linked to the receptor genes. It is suggested that the activator RNA produced by a single integrator gene could act upon a number of different receptor genes. In this way a single integrator could call forth the synthesis of a whole group of different enzymes. For example the synthesis of all of the enzymes of the urea cycle could be initiated simultaneously.

Just as the receptor gene is supposed to regulate its adjacent producer gene, so the integrator gene is regarded as being controlled by an adjacent **sensor gene**. This is a sequence of DNA which responds to an external agent such as a hormone or an inducer. Again the details of the interaction between the gene and the external agent are not an essential part of the model.

With these basic features of the model in mind (they are summarized in Fig. 50) we can begin to build up the degree of complexity which Britten and Davidson envisaged in the operation of the various elements. Firstly, a producer gene may have more than one receptor so that it can respond to different stimuli via the action of different sensor and integrator genes (Case II in Fig. 51). This would explain how a single enzyme could be induced by more than one substance, as liver tyrosine aminotransferase is induced by cortisone or by adrenalin. Again, a single integrator gene might act on more than one receptor gene. Such a scheme (Case III in Fig. 51) would explain the way in which more than one gene may be called upon to act by a single stimulus as are the genes of the urea cycle enzymes during amphibian metamorphosis.

A further level of complexity becomes possible if it is postulated that a single sensor gene may have more than one integrator gene. Case IV shows a very simple example of such regulation, in which producer genes A and B would respond to stimuli affecting sensor 1, while producer genes A and C would respond to stimuli affecting sensor 2. By combining a multiplicity of integrators attached to each sensor

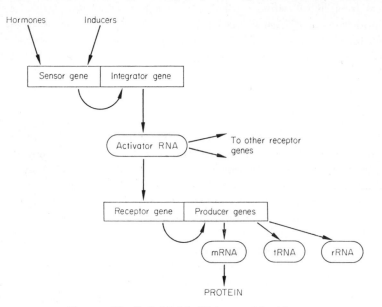

FIG. 50. The Basic Model of Britten and Davidson

and of receptors attached to each producer a vast degree of complexity becomes possible. A simple example of such a system is given in Case V. Here, application of stimulus 1 would bring about activity of producers A and B; stimulus 2 would activate producers A, B and C, and stimulus 3 would activate producers B, C and F. The reader should be able to envisage much more complex networks of interaction and reference to the paper of Britten and Davidson will provide more examples.

A striking feature of this theory is that for each sequence of DNA which produces RNA entering the cytoplasm there are considerable stretches of DNA which produce RNA molecules functioning within the confines of the nucleus. The producer genes represent only a relatively small part of the total DNA of the genome, and so the relatively large amount of DNA in the nuclei of eukaryotes may be explained to some extent. The activator RNA might be identical with a part of the so-called nuclear heterogeneous RNA which is synthesized rapidly but which does not pass through the nuclear membrane, being broken down within the nucleus.

DNA-RNA hybridization studies carried out by McCarthy and Shearer [121] showed that there were in the RNA confined to the nucleus sequences of bases which were absent from the cytoplasm. The conditions of these hybridization experiments, using short annealing times and low nucleic acid concentrations, would have caused the hybridization of only those most frequently occurring sequences. It is presumed that among the activator RNA molecules there will be substantial similarities of base sequence and Britten and Davidson regard the RNA confined to the nucleus and hybridizing under the conditions of McCarthy and Shearer as probably consisting of activator RNA. Moreover, they would argue that those fractions of RNA which

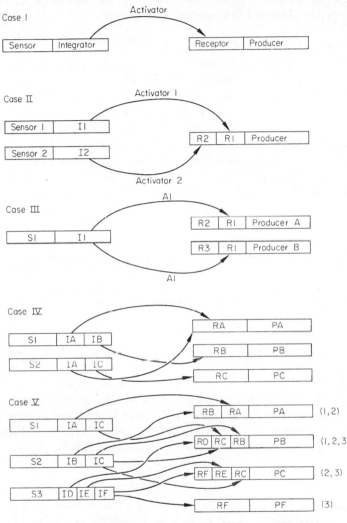

FIG. 51. Variations on the Basic Model of Britten and Davidson

undergo striking changes during differentiation as revealed by other hybridization experiments [131, 33] described in Chapter 5 as also probably composed to a large extent of repetitious RNA which could be activator RNA.

This model of gene regulation provides a good explanation of how a single stimulus can bring forth many different responses, and of how similar responses can be brought about by different stimuli. The way in which different patterns of enzymes can be found in different tissues is also rather well accounted for by the more complex networks of interaction between sensors, integrators and producers. It does not appear to provide any mechanism to explain the intrinsic regulations of cellular differentiation,

such as the persistence of determination or differentiation through many cell generations. Nor does the model provide explanations of the different time scales of events
in cellular differentiation. Further consideration of the theory will be found in a
later paper by Davidson and Britten [31].

Crick's model of the chromosome

The basic ideas of Britten and Davidson's theory have been taken by Crick and used
in an attempt to provide a model of the structure of the chromosome of higher
organisms [30]. Crick distinguishes between the DNA which is involved in coding
for polypeptides, that is the producer genes of Britten and Davidson, and the DNA
which has a control function, representing the sensor, integrator and receptor genes.
Crick postulates that the physical states of these types of DNA are different, the
'coding' or producer DNA being fibrous and the **control** DNA having a globular
structure. This globular DNA may have some regions in which the strands are not
held together by base-pairing, since Crick believes that the recognition of base
sequences by controlling proteins is more likely when the protein can interact closely
with the bases of the single-stranded form. The single-stranded state may be maintained by interaction with proteins such as histones. Crick relates his postulates of
DNA structure to the banded appearance of polytene chromosomes by supposing
that the control elements are in the bands of the chromosomes, while the coding
elements form the interband regions. The band regions are more extensive than the
interbands and it is supposed that control elements contain more DNA than coding
elements. This is in accord with the theory of Britten and Davidson and with the
observation that the DNA content of higher organisms is much higher than that
needed to code for the number of different proteins that are currently believed to be
synthesized.

The theory of Britten and Davidson, and Crick's extension of it, have been discussed at some length here, but the reader should realize that other models have been
put forward which merit consideration. In a short book of this kind it would be
inappropriate to devote too much space to the discussion of hypothetical systems, but
the interested reader may care to read the papers of Georgiev [42, 43] and of Paul [99]
which offer further proposals.

The cell cycle, cell division and differentiation

The influence of hormones and embryonic inducers in cell differentiation is well
established, and in one case that has been studied in detail, the induction of liver
tyrosine aminotransferase by corticosteroids, it appears that the cells are sensitive to
the inducing agent only at certain stages in the cell cycle. In some other differentiating
cell systems the external stimulus appears to trigger cell division before differentiation
takes place. This appears to be the case with the differentiating pancreas rudiment
[48], and it is a sufficiently common phenomenon that Holtzer has coined the phrase
quantal mitosis for such a division preceding differentiation.

The question of an antagonism between cell division and differentiation has
been much discussed in the past, but it is now clear that at least some differentiated
cells are capable of division and that other cells divide while in the determined, if

not the overtly differentiated, state. Nevertheless, there is some evidence that at certain stages of cell division the nucleus is susceptible to cytoplasmic influences which are ineffective at other stages. Gurdon and Woodland [54] report that cytoplasmic proteins can interact with chromosomes during telophase of mitosis, and that they are dissociated from the chromosomes at the following mitosis. It may be that mitosis, with its attendant breakdown of the nuclear membrane, permits the genome to develop new associations with informational molecules, so that new patterns of genetic activity can be programmed. Gurdon and Woodland suggest that during cell division some genes might synthesize products which reprogramme the genome so that it continues the same pattern of activity after mitosis as before, unless there is some specific interference from other stimuli. This would then provide a mechanism for the quasi-hereditary nature of differentiation, the so-called 'stability of the differentiated state'. An alternative explanation might be called for in some cases since the external stimulus might be necessary as a continuing presence if the differentiation is to remain stable. This could be the case with the cornea of the toad *Xenopus laevis* which, in the absence of a lens, may dedifferentiate and then give rise to a lens (see Chapter 6). Certainly, if we allow for either positive or negative interference in the nucleus from the cytoplasm, the proposal of Gurdon and Woodland appears to provide an adequate explanation for the inheritance of the state of differentiation or determination.

The few theories mentioned in this chapter must be understood as attempts to bring together our present very limited knowledge of the molecular control of cytodifferentiation. More knowledge is being gained at a rapid rate and we cannot expect that hypotheses proposed at present will stand for long without modification. On the other hand the experimental data outlined in earlier parts of the book, and the great mass of observations on cytodifferentiation which have been excluded from a book of this size, must all be accommodated in any theory which is to be convincing. Much more experimental work will be done and some of our present data may eventually appear trivial. Even so, though hypotheses and theories may come and go, sound experimental observations must stand as the foundation stones upon which the superstructure of theory can be built.

GLOSSARY

Acidic proteins of the chromatin Proteins present in chromatin which are not histones and which lack the high content of basic amino acids which characterize the histones. They are not particularly acidic in composition and are, perhaps, better known as 'non-histone proteins'.

Actinomycin D An antibiotic which inhibits RNA synthesis by binding to DNA and impeding the action of DNA-dependent RNA polymerase. This may not be the only action of the drug.

Activator RNA A postulate of Britten and Davidson. See p. 101.

Allosteric effector A substance which binds stereospecifically to an enzyme at a site distinct from the active centre and which affects the catalytic capability of the enzyme either by inhibition or enchancement.

Aneuploid See under polyploid.

Apoprotein That portion of a conjugate protein molecule which consists entirely of polypeptide material and does not include the prosthetic groups. Similarly, an **apoenzyme** is the polypeptide portion of an enzyme, excluding prosthetic groups.

Balbiani ring An RNA chromosome puff of a particularly large size.

Band—of Polytene Chromosome That part of a polytene chromosome which readily stains with the Feulgen reagents.

Blastema A mass of undifferentiated cells, the differentiation of which gives rise to a tissue or organ. The formation of a blastema is an important stage in regeneration of organs.

Chondrocyte One of the cells which form cartilage tissue.

Chondrogenesis The process by which cartilage is formed.

Chromatin The substance of which chromosomes are formed, containing DNA, histones, non-histone proteins (acidic proteins) and perhaps RNA. Chromatin exists in two forms: **heterochromatin** takes up electron-dense stains and is believed to be generally inactive in transcription while **euchromatin** can be transcribed to produce RNA. Some of the heterochromatin (facultative heterochromatin) can revert to euchromatin while constitutive heterochromatin does not so revert.

Chromosome puffs Regions of temporary expansion at particular sites on a polytene chromosome. RNA puffs, the most common type, are regions with enhanced RNA production in which the DNA may be less condensed, so occupying a larger volume but without being increased in quantity. DNA puffs are found in some species of Diptera, such as *Rhynchosciara*, and involve an increase in the quantity of DNA over that normally present in that region of the chromosome.

Clone A population of cells derived by mitosis from a single cell and hence of identical genotype. Also used of a group of organisms derived by asexual reproduction from a single parent and so genetically identical.

Coding DNA In the chromosome model of Crick, that DNA which acts as the template for messenger or other cytoplasmic RNA, such as transfer or ribosomal RNA. Equivalent to the producer gene of Britten and Davidson.

Control DNA In Crick's chromosome model, that DNA which regulates the activity of coding DNA. It would represent the receptor, integrator and sensor genes of Britten and Davidson.

Crystallin One of a group of structural proteins found in the lens of the eye.

Cytodifferentiation The process by which cells of different type acquire their differences.

Dedifferentiation The process whereby differentiated, specialized cells lose some of their specific structure and function. This may precede a phase of cell multiplication by mitosis and subsequent differentiation either to the original cell type or to a different type of cell. Dedifferentiation can be an important phase of regeneration.

Determination The process by which a tissue or cell comes to have only one possible fate in embryonic development; when no experimental manipulation can be shown to direct it into alternative channels.

Effector That substance which combines specifically with the repressor. In inducible enzyme systems it is the inducer; in repressible systems it is the inhibitor.

Erythropoiesis The process by which new erythrocytes are formed.

Erythropoietin A substance which stimulates the production of erythrocytes.

Eukaryotes The higher organisms, characterized by a distinct nucleus surrounded, except during nuclear division, by a nuclear membrane.

Functional gene unit In higher organisms, a group of enzymes, the synthesis of which responds in a similar manner to a particular stimulus.

Gastrulation The process in embryonic development by which an essentially single-layered structure, the blastula, is turned into a multi-layered structure, the gastrula. The process may involve folding movements, cell migration, cell division or all of these processes.

Genome The entire range of different genes to be found in an individual organism.

Genotype The genetic constitution of an organism as distinct from the characteristics made manifest in the phenotype.

Germ line Those cells of which the progeny are able to give rise to gametes.

Haemocytoblast An early stage in the differentiation of the erythrocyte. The cell has a large nucleus and an abundance of ribosomes, but no haemoglobin.

Hepatocyte The most abundant type of cell in the liver.

Histones A group of proteins with a characteristically high content of basic amino acids such as lysine and arginine, which form a major constituent of eukaryotic chromosomes.

Holoprotein The entire conjugate protein molecule, consisting of the apoprotein, or polypeptide portion, together with any prosthetic groups. Similarly, a **holoenzyme** is the entire enzyme with its prosthetic groups.

Inducer—genetic A genetic inducer is a substance which on entering a cell brings about the synthesis of an inducible enzyme. In bacterial enzyme induction it is usually the substrate of the enzyme or an analogue of the substrate.

Inducible enzyme An enzyme which is not normally synthesized by a cell but which can be synthesized following treatment with an inducer.

Initiation factor A protein which acts catalytically in protein synthesis by stimulating the commencement of translation of the messenger RNA molecule.

Integrator gene Postulate of Britten and Davidson, see p. 101.

Interband—of Polytene Chromosome The region between the bands and in which the DNA is not readily stained.

Isoenzyme or isozyme One of a group of enzymes of differing molecular properties which nevertheless catalyse identical chemical reactions.

Karyotype The description of the morphology of all of a set of chromosomes in an organism or cell.

Lens fibre An elongated type of cell forming the bulk of the lens of the eye, possessing a high concentration of structural proteins (crystallins) and lacking nucleus and most cell organelles.

Lens placode The thickened region of the ectoderm which gives rise to the lens vesicle and hence to the lens of the eye.

Meiosis The form of nuclear division which results in daughter nuclei with half the chromosome content of the parent nucleus.

Metaplasia The transformation of one type of differentiated tissue into another type.

Microfilaments Elongated structures within the cytoplasm with a diameter of about 5 nm, which are composed of a protein which may be related to actin.

Microtubules Elongated, hollow structures within cytoplasm with a diameter of about 23 nm, built up from the protein tubulin.

Mitosis The form of nuclear division which leads to daughter cells with an identical chromosome complement to that of the parent nucleus.

Modulation A short-term difference between cells, being readily reversible and possibly not requiring the expression of any characteristic pattern of genes.

Myoblast One of the uninucleate cells which by a process of cell fusion may give rise to the myotube which ultimately differentiates into muscle tissue.

Neural plate That region of the ectoderm which gives rise to the neural tube by the infolding movements of neurulation.

Neurulation The embryonic process whereby the neural tube is formed, chiefly by infolding of the dorsal ectoderm of the gastrula, followed by tissue fusion.

Nuclear heterogeneous RNA A class of RNA molecules of large but variable size which is confined to the cell nucleus and which probably gives rise to messenger RNA by a controlled process of molecular breakdown.

Nucleolar organizer The part of a chromosome which gives rise to the nucleolus in the interphase nucleus. It is the region of the chromosome that contains the DNA which acts as the template for ribosomal RNA.

Operator gene In the operon theory of Jacob and Monod (see p. 58), a gene which regulates the activity of one or more structural genes, being itself under the control of the regulator gene.

Operon The group of structural genes and their operator and regulator genes which forms the functional unit in the theory of Jacob and Monod (see p. 58).

Phenotype The observable form of an organism resulting from the interaction of its genotype and environment during development.

Polycistronic messenger A single molecule of messenger RNA which serves as the template for more than one type of protein.

Polyploid Possessing more than two complete sets of chromosomes. In polyploidy the number of chromosomes is the haploid number multiplied by an integral number. In aneuploidy there is a departure from an integral multiple of the haploid number of chromosomes, as when there is one more or one less than the usual number of an individual chromosome.

Polysome, polyribosome An assembly of ribosomes and messenger RNA which is the functioning unit in protein synthesis.

Polytene chromosome A form of interphase chromosome in which the basic chromatid has replicated to give many identical copies which remain held together in parallel array.

Producer gene Term used in the model of Britten and Davidson to describe a gene which codes for either messenger RNA or any other class of cytoplasmic RNA such as transfer or ribosomal RNA. Equivalent to the coding DNA of Crick's model and similar to the structural gene of Jacob and Monod.

Prokaryotes The lower organisms, i.e. bacteria and blue-green algae, lacking a nucleus with a nuclear membrane.

Prosthetic group That portion of a protein molecule which is not polypeptide in nature, e.g. the haem group of haemoglobin or the polysaccharide side chain of a mucopolysaccharide.

Quantal mitosis Term proposed by Holtzer for a mitotic division which ends a phase of cell proliferation and initiates terminal differentiation of cells.

Receptor gene Postulate of Britten and Davidson, see p. 101.

Regulator gene In the operon theory of Jacob and Monod (see p. 58), the gene responsible for the synthesis of the repressor substance.

Repressor A product of the regulator gene which interacts with effector substances and controls the activity of the operator gene. In inducible enzyme systems the repressor combines with the inducer and then is no longer able to bind to and inactivate the operator gene. In repressible systems, the repressor substance only binds to and inactivates the operator after it has combined with the effector. The repressor is a protein.

Sensor gene Postulate of Britten and Davidson, see p. 101.

Somatic cell Those cells of the body which are neither gametes nor give rise to gametes.

Stem cell One of a population of cells which undergoes repeated mitosis, giving rise at each division to another stem cell and a cell which ultimately undergoes terminal differentiation.

Structural gene A sequence of DNA which specifies the amino acid sequence of the polypeptide chain of an enzyme or other protein. The term is used in connexion with the operon theory of Jacob and Monod (see p. 58), but corresponds to the producer gene of Britten and Davidson.

Syncytium A form of organization of tissue in which the cytoplasm is not divided into separate mononucleate cells but in which many nuclei may be present without dividing cell membranes.

Synergism The phenomenon by which the effect of two or more agents on a system is greater than the sum of the effects of each agent when acting alone.

Terminal differentiation The state of cytodifferentiation at which no more differentiation is possible and in which cell division cannot occur.

Tissue-specific Restricted in distribution to the cells of a particular tissue but perhaps common to that tissue in many species.

Totipotent Possessing the ability to give rise to any of the cell types normally present in the body.

Transcription The process by which DNA acts as a template specifying the base sequence of RNA as it is synthesized. In the living cell one strand of the DNA acts as a template while the complementary strand is not transcribed.

Translation The process by which RNA acts as a template specifying the amino acid sequence of a protein as it is synthesized.

Zygote The single diploid cell formed after fertilization and the fusion of the male and female nuclei.

Zymogen The inactive precursor of an enzyme, e.g. trypsinogen, which gives rise to trypsin. Also used of the particles visible in secretory cells which consist of concentrations of these enzyme precursors.

REFERENCES

1   ALESCIO T., MOSCONA M. & MOSCONA A.A. (1970) Induction of glutamine synthetase in embryonic retina. *Experimental Cell Research*, **61**, 342–6.

2   ARMS K. (1968) Cytonucleoproteins in cleaving eggs of *Xenopus laevis*. *Journal of Embryology and Experimental Morphology*, **20**, 367–74.

3   ASHBURNER M. (1967) Patterns of puffing activity in the salivary gland chromosomes of *Drosophila*. *Chromosoma*, **21**, 398–428.

4   BÄCKSTRÖM S. (1963) 6-Phosphogluconate dehydrogenase in sea urchin embryos. *Experimental Cell Research*, **32**, 566–9.

5   BALINSKY B.I. (1970) *An Introduction to Embryology*, 3rd ed. Philadelphia: Saunders.

6   BEATTIE D.S., BASFORD R.E. & KORITZ S.B. (1966) Studies on the biosynthesis of mitochondrial protein components. *Biochemistry*, **5**, 926–30.

7   BEERMANN W. (1952) Chromomerenkonstanz und spezifische Modifikationen der Chromosomenstruktur in der Entwicklung und Organdifferenzerung von *Chironomus tentans*. *Chromosoma*, **5**, 139–98.

8   BEERMANN W. (1956) Nuclear differentiation and functional morphology of chromosomes. *Cold Spring Harbor Symposia on Quantitative Biology*, **21**, 217–30.

9   BEERMANN W. (1964) Control of differentiation at the chromosomal level. *Journal of Experimental Zoology*, **157**, 49–61.

10 BIRNSTIEL M.L., CHIPCHASE M. & SPEIRS J. (1971) The ribosomal RNA cistrons. *Progress in Nucleic Acid Research and Molecular Biology*, **11**, 351–89.

11 BONNER J., DAHMUS M.E., FAMBROUGH D., HUANG R.-C.C., MARUSHIGE K. & TUAN D.Y.H. (1968) The biology of isolated chromatin. *Science*, **159**, 47–56.

12 BORST P. & KROON A.M. (1969) Mitochondrial DNA: physicochemical properties, replication and genetic function. *International Review of Cytology*, **26**, 108–90.

13 BOVERI T. (1899) Die Entwicklung von Ascaris megalocephala mit besonderer Rücksicht auf die Kernverhältnisse. *Festchrift F.C. von Kupffer*. Jena.

14 BRACHET J. (1960) *The Biochemistry of Development*. London: Pergamon.

15 BRAUN A.C. (1959) A demonstration of the recovery of the crown-gall tumor cell with the use of complex tumors of single-cell origin. *Proceedings of the National Academy of Sciences of the United States of America*, **45**, 932–8.

16 BREUER M.E. & PAVAN C. (1955) Behavior of polytene chromosomes of Rhynchosciara angelae at different stages of larval development. *Chromosoma*, **7**, 371–86.

17 BRITTEN R.J. & DAVIDSON E.H. (1969) Gene regulation for higher cells: a theory. *Science*, **165**, 349–57.

18 BRITTEN R.J. & KOHNE D.E. (1968) Repeated sequences in DNA. *Science*, **161**, 529–40.

19 BROWN D.D. & LITTNA E. (1964) RNA synthesis during the development of *Xenopus laevis*, the South African clawed toad. *Journal of Molecular Biology*, **8**, 669–87.

20 BURNS R.K. (1955) Experimental reversal of sex in the gonads of the opossum *Didelphis vir-*

giniana. Proceedings of the National Academy of Sciences of the United States of America, **41,** 669–76.

21 CAHN R.D. & CAHN M.B. (1966) Heritability of cellular differentiation: clonal growth and expression of differentiation in retinal pigment cells in vitro. *Proceedings of the National Academy of Sciences of the United States of America,* **55,** 106–13.

22 CAMPBELL J.C. & JONES K.W. (1968) The *in vitro* development of lens from cornea of larval *Xenopus laevis. Developmental Biology,* **17,** 1–15.

23 CHANCE R.E., ELLIS R.M. & BROMER W.W. (1968) Porcine proinsulin: characterization and amino acid sequence. *Science,* **161,** 165–7.

24 CLAYTON R.M. (1970) Problems of differentiation in the vertebrate lens. *Current Topics in Developmental Biology,* **5,** 115–80.

25 CLAYTON R.M., TRUMAN D.E.S. & CAMPBELL J.C. (1972) A method for direct assay of messenger RNA turnover for different crystallins in the chick lens. *Cell Differentiation,* **1,** 25–35.

26 CLEVER U. (1968) Regulation of chromosome function. *Annual Review of Genetics,* **2,** 11–30.

27 COHEN P.P. (1970) Biochemical differentiation during amphibian metamorphosis. *Science,* **168,** 533–43.

28 COON H.G. (1966) Clonal stability and phenotypic expression in chick cartilage cells *in vitro. Proceedings of the National Academy of Sciences of the United States of America,* **55,** 66–73.

29 COULOMBRE J.L. & COULOMBRE A.J. (1971) Metaplastic induction of scales and feathers in the corneal anterior epithelium of the chick embryo. *Developmental Biology,* **25,** 464–78.

30 CRICK F.H.C. (1971) General model for the chromosomes of higher organisms. *Nature,* **234,** 25–7.

31 DAVIDSON E.H. & BRITTEN R.J. (1971) Note on the control of gene expression during development. *Journal of Theoretical Biology,* **32,** 123–30.

32 DAVIDSON R.G. & CORTNER J.A. (1967) Mitochondrial malate dehydrogenase: a new genetic polymorphism in man. *Science,* **157,** 1569–71.

33 DENIS H. (1968) Role of messenger ribonucleic acid in embryonic development. *Advances in Morphogenesis,* **7,** 115–50.

34 DI BERARDINO M. & HOFFNER N. (1970) Origin of chromosomal abnormalities in nuclear transplants—a reevaluation of nuclear differentiation and nuclear equivalence in amphibians. *Developmental Biology,* **23,** 185–209.

35 DIXON M. & WEBB E.C. (1964) *The Enzymes,* 2nd ed. London: Longmans.

36 EGUCHI G. (1963) Electron microscope studies on lens regeneration. I. Mechanism of depigmentation of the iris. *Embryologia,* **8,** 45–62.

37 EGUCHI G. (1964) Electron microscope studies on lens regeneration. II. Formation and growth of lens vesicle and differentiation of lens fibers. *Embryologia,* **8,** 247–87.

38 FAWCETT D.W. (1966) *The Cell, its Organelles and Inclusions.* Philadelphia: Saunders.

39 FLICKINGER R.A., COWARD S.J., MIYAGI M., MOSER C. & ROLLINS E. (1965) The ability of DNA and chromatin of developing frog embryo to prime for RNA polymerase-dependent RNA synthesis. *Proceedings of the National Academy of Sciences of the United States of America,* **53,** 783–90.

40 GALL J.G. (1969) The genes for ribosomal RNA during oogenesis. *Genetics,* **61,** Supplement 121–32.

41 GARREN L.D., HOWELL R.R., TOMKINS G.M. & CROCCO R.M. (1964) A paradoxical effect of actinomycin D: the mechanism of regulation of enzyme synthesis by hydrocortisone. *Proceedings of the National Academy of Sciences of the United States of America,* **52,** 1121–9.

42 GEORGIEV G.P. (1969) On the structural organization of operon and the regulation of RNA synthesis in animal cells. *Journal of Theoretical Biology,* **25,** 473–90.

43 GEORGIEV G.P., RYSKOV A.P., COUTELLE C., MANTIEVA V.L. & AVAKYAN E.R. (1972) On the structure of transcription unit in mammalian cells. *Biochimica et Biophysica Acta,* **259,** 259–83.

44 GOLDWASSER E. (1966) Biochemical control of erythroid cell development. *Current Topics in Developmental Biology,* **1,** 173–211.

45 GRANICK S. & GIBOR A. (1967) The DNA of chloroplasts, mitochondria and centrioles. *Progress in Nucleic Acid Research and Molecular Biology,* **6,** 143–86.

46 GREENGARD O. (1971) Enzymic differentiation in mammalian liver. *Essays in Biochemistry*, 7, 159–205.

47 GROBSTEIN C. (1959) Differentiation of vertebrate cells. In *The Cell*, ed. BRACHET J. & MIRSKY A. Vol. 1, 437–96. New York: Academic Press.

48 GROBSTEIN C. (1964) Cytodifferentiation and its control. *Science*, 143, 643–50.

49 GROUSE L., CHILTON M.-D., & McCARTHY B.J. (1972) Hybridization of ribonucleic acid with unique sequences of mouse deoxyribonucleic acid. *Biochemistry*, 11, 798–805.

50 GURDON J.B. (1962) The developmental capacity of nuclei taken from intestinal epithelium cells of feeding tadpoles. *Journal of Embryology and Experimental Morphology*, 10, 622–40.

51 GURDON J.B. (1964) The transplantation of living cell nuclei. *Advances in Morphogenesis*, 4, 1–43.

52 GURDON J.B. (1968) Nucleic acid synthesis in embryos and its bearing on cell differentiation. *Essays in Biochemistry* 4, 25–68.

53 GURDON J.B. & UEHLINGER V. (1966) 'Fertile' intestine nuclei. *Nature*, 210, 1240–1.

54 GURDON J.B. & WOODLAND H.R. (1970) On the long-term control of nuclear activity during cell differentiation. *Current Topics in Developmental Biology*, 5, 39–70.

55 HAM A.W. (1969) *Histology*, 6th ed. London: Pitman.

56 HAMBURGH M. (1971) *Theories of Differentiation*. London: Arnold.

57 HAMILTON T.H. (1968) Control by estrogen of genetic transcription and translation. *Science*, 161, 649–61.

58 HARRIS H. (1970) *Cell Fusion*. Oxford: Clarendon Press.

59 HARRIS H. (1970) *The Principles of Human Biochemical Genetics*. Amsterdam: North Holland.

60 HAUSCHKA S.D. (1968) Clonal aspects of muscle development and the stability of the differentiated state. In *The Stability of the Differentiated State*, ed. Ursprung H. 37–57. Berlin: Springer-Verlag.

61 HELMREICH E. (1969) Control of synthesis and breakdown of glycogen, starch and cellulose. In *Comprehensive Biochemistry*, ed. Florkin M. & Stotz E.H. Vol. 17, 17–92. Amsterdam: Elsevier.

62 HEYWOOD S.M., DOWBEN R.M. & RICH A. (1967) The identification of polysomes synthesising myosin. *Proceedings of the National Academy of Sciences of the United States of America*, 57, 1002–9.

63 HORSTADIUS S. & WOLSKY A. (1936) Studien über die Determination der Bilateralsymmetrie des jungen Seeigelkeimes. *Wilhelm Roux Archiv für Enwicklungsmechanik der Organismen*, 135, 69–113.

64 HUEHNS E.R., DANCE N., BEAVEN G.H., HECHT F. & MOTULSKY A.G. (1964) Human embryonic hemoglobins. *Cold Spring Harbor Symposia on Quantitative Biology*, 29, 327–31.

65 ILAN J. (1969) The role of tRNA in translational control of specific mRNA during insect metamorphosis. *Cold Spring Harbor Symposia on Quantitative Biology*, 34, 787–91.

66 INGRAM V.M. (1967) On the biosynthesis of hemoglobin. *Harvey Lectures*, 61, 43–70.

67 JACOB F. & MONOD J. (1961) Genetic regulatory mechanisms in the synthesis of proteins. *Journal of Molecular Biology*, 3, 318–56.

68 JANAKI-AMMAL E.K. (1940) Chromosome diminution in a plant. *Nature*, 146, 839–40.

69 KADENBACH B. (1967) Synthesis of mitochondrial proteins. The synthesis of cytochrome c *in vitro*. *Biochimica et Biophysica Acta*, 138, 651–4.

70 KAMIYAMA M. & WANG T.Y. (1971) Activated transcription from rat liver chromatin by non-histone proteins. *Biochimica et Biophysica Acta*, 228, 563–76.

71 KING T.J. & BRIGGS R. (1956) Serial transplantation of embryonic nuclei. *Cold Spring Harbor Symposia on Quantitative Biology*, 21, 271–90.

72 KNOX W.E. & GREENGARD O. (1965) The regulation of some enzymes of nitrogen metabolism—an introduction to enzyme physiology. *Advances in Enzyme Regulation*, 3, 247–313.

73 KONIGSBERG I.R. (1963) Clonal analysis of myogenesis. *Science*, 140, 1273–84.

74 LEVERE R.D., KAPPAS A. & GRANICK S. (1967) Stimulation of hemoglobin synthesis in chick blastoderms by certain 5β androstane and 5β pregnane steroids. *Proceedings of the National Academy of Sciences of the United States of America*, 58, 985–90.

75 LEWIN B.M. (1970) *The Molecular Basis of Gene Expression*. London: Wiley-Interscience.

76 LIAO S., BARTON R.W. & LIN A.H. (1966) Differential synthesis of ribonucleic acid in prostatic

nuclei: evidence for selective gene transcription induced by androgens. *Proceedings of the National Academy of Sciences of the United States of America*, 55, 1593–600.

77   LITTAU V.C., ALLFREY V.G., FRENSTER J.H. & MIRSKY A.E. (1964) Active and inactive regions of nuclear chromatin as revealed by electron microscope autoradiography. *Proceedings of the National Academy of Sciences of the United States of America*, 52, 93–100.

78   MACGILLIVRAY A.J., CARROLL D. & PAUL J. (1971) The heterogeneity of the non-histone chromatin proteins from mouse tissues. *FEBS Letters*, 13, 204–8.

79   MARKERT C.L. (1964) Developmental genetics. *Harvey Lectures*, 59, 187–218.

80   MARKERT C.L. & URSPRUNG H. (1962) The ontogeny of isozyme patterns of lactate dehydrogenase in the mouse. *Developmental Biology*, 5, 363–81.

81   MARKERT C.L. & WHITT G.S. (1968) Molecular varieties of isoenzymes. *Experientia*, 24, 977–91.

82   MARKS P.A. & KOVACH J.S. (1966) Development of mammalian erythroid cells. *Current Topics in Developmental Biology*, 1, 213–52.

83   McCARTHY B.J. & HOYER B.H. (1964) Identity of DNA and diversity of messenger RNA molecules in normal mouse tissues. *Proceedings of the National Academy of Sciences of the United States of America*, 52, 915–22.

84   MEDOFF J. (1967) Enzymatic events during cartilage differentiation in the chick embryonic limb bud. *Developmental Biology*, 16, 118–43.

85   MELLI M. & PEMBERTON R.E. (1972) A new method of studying the precursor-product relationship between high molecular weight RNA and messenger RNA. *Nature New Biology*, 236, 172–4.

86   METZ C.W. (1938) Chromosome behaviour, inheritance and sex determination in Sciara. *American Naturalist*, 72, 485–520.

87   MONOD J., CHANGEUX J.-P. & JACOB F. (1963) Allosteric proteins and cellular control systems. *Journal of Molecular Biology*, 6, 306–29.

88   MOOG F. (1965) Enzyme development in relation to functional differentiation. In *The Biochemistry of Animal Development*, ed. Weber R. Vol. 1, 307–65. New York: Academic Press.

89   MORGAN T.H. (1934) *Embryology and Genetics*. New York: Columbia University Press.

90   MOSS B. & INGRAM V.M. (1968) Hemoglobin synthesis during amphibian metamorphosis. II. Synthesis of adult hemoglobin following thyroxin administration. *Journal of Molecular Biology*, 32, 493–502.

91   NASS M.M.K. & BUCK C.A. (1970) Studies on mitochondrial tRNA from animal cells. II. Hybridization of aminoacyl-tRNA from rat liver mitochondria with heavy and light complementary strands of mitochondrial DNA. *Journal of Molecular Biology*, 54, 187–98.

92   NEEDHAM J. (1942) *Biochemistry and Morphogenesis*. Cambridge: University Press.

93   NEUTRA M. & LEBLOND C.P. (1969) The Golgi apparatus. *Scientific American*, 220, 100–7.

94   NEWSHOLME E.A. (1970) Theoretical and experimental considerations on the control of glycolysis in muscle. In *Essays in Cell Metabolism*, ed. Bartley W., Kornberg H.L. & Quayle J.R. 189–223. London: Wiley.

95   OKADA T.S., EGUCHI G. & TAKEICHI M. (1971) The expression of differentiation by chicken lens epithelium in *in vitro* cell culture. *Development, Growth and Differentiation*, 13, 323–36.

96   O'MALLEY B.W., ROSENFELD G.C., COMSTOCK J.P. & MEANS A.R. (1972) Steroid hormone induction of a specific translatable messenger RNA. *Nature New Biology*, 240, 45–8.

97   PAPACONSTANTINOU J. (1967) Molecular aspects of lens cell differentiation. *Science*, 156, 338–46.

98   PAUL J. (1970) DNA masking in mammalian chromatin: a molecular mechanism for determination of cell type. *Current Topics in Developmental Biology*, 5, 317–52.

99   PAUL J. (1972) General theory of chromosome structure and gene activation in eukaryotes. *Nature*, 238, 444–6.

100   PAUL J. & GILMOUR R.S. (1966) Restriction of deoxyribonucleic acid template activity in chromatin is organ-specific. *Nature*, 210, 992–3.

101   PAUL J. & GILMOUR R.S. (1968) Organ-specific restriction of transcription in mammalian chromatin. *Journal of Molecular Biology*, 34, 305–16.

102   PEARCE T.L. & ZWAAN J. (1970) A light and electron microscopic study of cell behavior and microtubules in the chicken lens using colcemid. *Journal of Embryology and Experimental Morphology*, 23, 491–507.

103 PERKOWSKA E., MACGREGOR H.C. & BIRNSTIEL M.L. (1968) Gene amplification in the oocyte nucleus of mutant and wild-type *Xenopus laevis*. *Nature*, **217**, 649–50.

104 PERRY M.M., JOHN H.A. & THOMAS N.S.T. (1971) Actin-like filaments in the cleavage furrow of newt egg. *Experimental Cell Research*, **65**, 249–53.

105 PERUTZ M.F. & LEHMANN H. (1968) Molecular pathology of human haemoglobin. *Nature*, **219**, 902–9.

106 PIATIGORSKY J., WEBSTER H. deF. & CRAIG S.P. (1972) Protein synthesis and ultrastructure during the formation of embryonic chick lens fibers *in vivo* and *in vitro*. *Developmental Biology* **27**, 176–89.

107 PITOT H.C., PERAINO C., LAMAR C. & KENNAN A.L. (1965) Template stability of some enzymes in rat liver and hepatoma. *Proceedings of the National Academy of Sciences of the United States of America*, **54**, 845–51.

108 RABINOVITZ M., FREEDMAN M.L., FISHER J.M. & MAXWELL C.R. (1969) Translational control in hemoglobin synthesis. *Cold Spring Harbor Symposia on Quantitative Biology*, **34**, 567–78.

109 RIFKIN M.R., WOOD D.D. & LUCK D.J.L. (1967) Ribosomal RNA and ribosomes from mitochondria of Neurospora crassa. *Proceedings of the National Academy of Sciences of the United States of America*, **58**, 1025–32.

110 RITOSSA F.M. & SPIEGELMAN S. (1965) Localization of DNA complementary to ribosomal RNA in the nucleolus organizer region of Drosophila melanogaster. *Proceedings of the National Academy of Sciences of the United States of America*, **53**, 737–45.

111 ROSENBERG M. (1971) Epigenetic control of lactate dehydrogenase subunit assembly. *Nature New Biology*, **230**, 12–14.

112 RUDKIN G.T. & CORLETTE S.L. (1957) Disproportionate synthesis of DNA in a polytene chromosome region. *Proceedings of the National Academy of Sciences of the United States of America*, **43**, 964–8.

113 RUTTER W.J., KEMP J.D., BRADSHAW W.S., CLARK W.R., RONZIO R.A. & SANDERS T.G. (1968) Regulation of specific protein synthesis in cytodifferentiation. *Journal of Cell Physiology*, **72**, supplement 1, 1–18.

114 SARGENT J.R. (1973) *Protein Metabolism in Animal Cells*. Oxford: Blackwell.

115 SAXEN L. & TOIVONEN S. (1962) *Primary Embryonic Induction*. London: Logos Press.

116 SCHERRER K. & MARCAUD L. (1968) Messenger RNA in avian erythroblasts at the transcriptional and translational levels and the problem of regulation in animal cells. *Journal of Cell Physiology*, **72**, supplement 1, 181–212.

117 SCHERRER K., SPOHR G., GRANBOULAN N., MOREL C., GROSCLAUDE J. & CHEZZI C. (1970) Nuclear and cytoplasmic messenger-like RNA and their relation to the active messenger RNA in polysomes of HeLa cells. *Cold Spring Harbor Symposia on Quantitative Biology*, **35**, 539–74.

118 SCHROEDER T.E. (1970) Neurulation in *Xenopus laevis*. An analysis and model based upon light and electron microscopy. *Journal of Embryology and Experimental Morphology*, **23**, 427–62.

119 SCOTT R.B. & BELL E. (1964) Protein synthesis during development: control through messenger RNA. *Science* **145**, 711–14.

120 SCRUTTON M.C. & UTTER M.F. (1968) The regulation of glycolysis and gluconeogenesis in animal tissues. *Annual Review of Biochemistry*, **37**, 249–302.

121 SHEARER R.W. & MCCARTHY B.J. (1967) Evidence for ribonucleic acid molecules restricted to the cell nucleus. *Biochemistry*, **6**, 283–9.

122 SMITH K.D., CHURCH R.B. & MCCARTHY B.J. (1969) Template specificity of isolated chromatin. *Biochemistry*, **8**, 4271–7.

123 SPELSBERG T.C., HNILICA L.S. & ANSEVIN A.T. (1971) Proteins of chromatin in template restriction. III. The macromolecules in specific restriction of the chromatin DNA. *Biochimica et Biophysica Acta*, **228**, 550–62.

124 SPEMANN H. (1928) Die Entwicklung seitlicher und dorso-ventraler Keimhälften bei verzögerter Kernversorgung. *Zeitschrift für wissenschaftliche Zoologie*, **132**, 105–34.

125 SPIEGELMAN S. (1948) Differentiation as the controlled production of unique enzymatic patterns. *Symposia of the Society for Experimental Biology*, **2**, 286–325.

126 STEDMAN E. & STEDMAN E. (1950) Cell specificity of histones. *Nature*, **166**, 780–1.

127 STEPHENS R.E. (1968) Reassociation of microtubule protein. *Journal of Molecular Biology*, **33**, 517–19.

128 STEWARD F.C. (1970) From cultured cells to whole plants: the induction and control of their growth and morphogenesis. *Proceedings of the Royal Society*, B, **175**, 1–30.

129 STEWARD F.C., MAPES M.O., KENT A.E. & HOLSTEN R.D. (1964) Growth and development of cultured plant cells. *Science*, **143**, 20–7.

130 SUEOKA N. & KANO-SUEOKA T. (1970) Transfer RNA and cell differentiation. *Progress in Nucleic Acid Research and Molecular Biology*, **10**, 23–53.

131 TAYLOR M.W., GRANGER G.A., BUCK C.A. & HOLLAND J.J. (1967) Similarities and differences among specific tRNA's in mammalian tissues. *Proceedings of the National Academy of Sciences of the United States of America*, **57**, 1712–19.

132 TENG C.-S. & HAMILTON T.H. (1969) Role of chromatin in estrogen action in the uterus, II. Hormone-induced synthesis of nonhistone acid proteins which restore histone-inhibited DNA-dependent RNA synthesis. *Proceedings of the National Academy of Sciences of the United States of America*, **63**, 465–72.

133 THORP F.K. & DORFMAN A. (1967) Differentiation of connective tissue. *Current Topics in Developmental Biology*, **2**, 151–90.

134 TIEDEMANN H. (1971) Extrinsic and intrinsic information transfer in early differentiation of amphibian embryos. *Symposia of the Society for Experimental Biology*, **25**, 223–34.

135 TILNEY L.G. (1968) The assembly of microtubules and their role in the development of cell form. *Developmental Biology*, supplement **2**, 63–102.

136 TJIO J.H. & PUCK T.T. (1958) The somatic chromosomes of man. *Proceedings of the National Academy of Sciences of the United States of America*, **44**, 1229–36.

137 TOMKINS G.M., GELHRTER T.D., GRANNER D., MARTIN D., SAMUELS H.H. & THOMPSON E.B. (1969) Control of specific gene expression in higher organisms. *Science*, **166**, 1474–80.

138 TOMKINS G.M., LEVINSON B.B., BAXTER J.D. & DETHLEFSEN L. (1972) Further evidence for posttranscriptional control of inducible tyrosine aminotransferase synthesis in cultured hepatoma cells. *Nature New Biology*, **239**, 9–14.

139 TRUMAN D.E.S. (1964) The fractionation of proteins from ox-heart mitochondria labelled *in vitro* with radioactive amino acids. *Biochemical Journal*, **91**, 59–64.

140 TRUMAN D.E.S., BROWN A.G. & CAMPBELL J.C. (1972) The relationship between the ontogeny of antigens and of the polypeptide chains of the crystallins during chick lens development. *Experimental Eye Research*, **13**, 58–69.

141 TRUMAN D.E.S., BROWN A.G. & RAO K.V. (1971) Estimation of the molecular weights of chick β- and δ-crystallins and their subunits by gel filtration. *Experimental Eye Research*, **12**, 304–10.

142 TRUMAN D.E.S. & KORNER A. (1962) Initial stages in the incorporation of amino acids into protein in rat-liver mitochondria. *Biochemical Journal*, **85**, 154–8.

143 TURKINGTON R.W. (1968) Hormone-dependent differentiation of mammary gland *in vitro*. *Current Topics in Developmental Biology*, **3**, 199–218.

144 URSPRUNG H., SMITH K.D., SOFER W.H. & SULLIVAN D.T. (1968) Assay systems for the study of gene function. *Science*, **160**, 1075–81.

145 VASIL V. & HILDERBRANDT A.C. (1965) Differentiation of tobacco plants from single, isolated cells in microcultures. *Science*, **150**, 890–2.

146 VENDRELY R. (1955) The deoxyribonucleic acid content of the nucleus. In *The Nucleic Acids*, ed. Chargaff E. & Davidson J.N. Vol. 2, 155–80. New York: Academic Press.

147 WADDINGTON C.H. (1956) *Principles of Embryology*. London: Allen & Unwin.

148 WALLACE H. & BIRNSTIEL M.L. (1966) Ribosomal cistrons and the nucleolus organiser. *Biochimica et Biophysica Acta*, **114**, 296–310.

149 WATSON J.D. (1970) *Molecular Biology of the Gene*, 2nd ed. New York: Benjamin.

150 WEBER G., SINGHAL R.L., STAMM N.B., LEA M.A. & FISHER E.A. (1966) Synchronous behavior pattern of key glycolytic enzymes: glucokinase, phosphofructokinase and pyruvate kinase. *Advances in Enzyme Regulation*, **4**, 59–81.

151 WEBER R. (1963) Behaviour and properties of acid hydrolases in regressing tails of tadpoles during spontaneous and induced metamorphosis *in vitro*. In *Lysosomes*, ed. De Reuck A.V.S. & Cameron M.P. 282–310. London: Churchill.

152 WEBER R. *The Biochemistry of Animal Development*. New York: Academic Press.
153 WEBER R. & BOELL E.J. (1962) Enzyme patterns in isolated mitochondria from embryonic and larval tissue of *Xenopus*. *Developmental Biology*, 4, 452–72.
154 WESSELS N.K., SPOONER B.S., ASH J.F., BRADLEY M.O., LUDUENA M.A., TAYLOR 'E.L., WRENN J.T. & YAMADA K.M. (1971) Microfilaments in cellular and developmental processes. *Science*, 171, 135–43.
155 WESSELS N.K. & WILT F.H. (1965) Action of actinomycin D on exocrine pancreas cell differentiation. *Journal of Molecular Biology*, 13, 767–71.
156 WILDE C.E. (1961) The differentiation of vertebrate pigment cells. *Advances in Morphogenesis*, 1, 267–300.
157 WILKIE D. (1964) *The Cytoplasm in Heredity*. London: Methuen.
158 WILLIAMSON R., CLAYTON R.M. & TRUMAN D.E.S. (1972) Isolation and identification of chick lens crystallin messenger RNA. *Biochemical and Biophysical Research Communications*, 46, 1936–43.
159 WILT F.H. (1967) The control of embryonic hemoglobin synthesis. *Advances in Morphogenesis*, 6, 89–125.
160 YAFFE D. (1969) Cellular aspects of muscle differentiation *in vitro*. *Current Topics in Developmental Biology*, 4, 37–77.

INDEX

Page numbers in **bold type** refer to the principal treatment of the topic concerned